Essentials of Environmental Toxicology

Essentials of Environmental Toxicology

The Effects of Environmentally Hazardous Substances on Human Health

W. William Hughes
Professor
School of Allied Health Professions
Associate Clinical Professor of Pathology and Human Anatomy
School of Medicine
Loma Linda University
Loma Linda, California

Taylor & Francis
Publishers since 1798

USA	Publishing Office:	Taylor & Francis
		325 Chestnut St., Suite 800
		Philadelphia, Pa 19106
		Tel: (215) 625-8900
		Fax: (215) 625-2940
	Distribution Center:	Taylor & Francis
		47 Runway Road, Suite G
		Levittown, Pa 19057
		Tel: (215) 269-0400
		Fax: (215) 269-0363
UK		Taylor & Francis Ltd.
		1 Gunpowder Square
		London EC4A 3DE
		Tel: 0171 583 0490
		Fax: 0171 583 0581

ESSENTIALS OF ENVIRONMENTAL TOXICOLOGY: The Effects of Environmentally Hazardous Substances on Human Health

2 3 4 5 6 7 8 9 0   MKMK   9 8 7 6

This book was set in Times Roman by Innodata Publishing Services Division. The editors were Catherine Simon, Sharon M. Twigg, and Elizabeth Dugger. Figure design by Louise Ceccarelli. Cover design by Michelle Fleitz. Interior book design by Bonny Gaston. Printing and binding by The Mack Printing Group.

A CIP catalog record for this book is available from the British Library.
∞ The paper in this publication meets the requirements of the ANSI Standard Z39.48-1984 (Permanence of Paper)

Library of Congress Cataloging-in-Publication Data

Hughes, W. William.
    Essentials of environmental toxicology: the effects of environmentally hazardous substances on human health/W. William Hughes.
        p. cm.

    1. Environmental toxicology.   I. Title.
RA1226.H84   1996   96-14283
615.9′02—dc20   CIP

ISBN 1-56032-469-4 (case)
ISBN 1-56032-470-8 (paper)

To Asa Thoresen
. . . a gentle and great man

Contents

Preface

The effects of environmentally hazardous substances on human health are an important and timely subject. My goal was to produce an introductory text similar in style to those found in the aged disciplines such as anatomy, biology, botany, and physiology—a text that emphasized the general state of environmental toxicology by concentrating on important principles and avoided the controversies inherent within any discipline (not majoring in the minors).

This text is written for students enrolled in environmental science and environmentally hazardous materials certificate, associate, and baccalaureate degree programs. It has been my experience, having taught environmental science and environmental toxicology during the past 16 years, that these students have diverse backgrounds. The text has been written with an awareness that many of you are just beginning post-high-school studies, some already have advanced degrees, and others are making career changes. It is assumed that the reader has a high-school familiarity with biology and chemistry, but no prior knowledge of environmental toxicology.

Essentials of Environmental Toxicology takes a superficial look at what in reality is a very large and complex body of knowledge. Any endeavor to distill a discipline into a few hundred pages of text is laden with an enormous responsibility to avoid misrepresenting data for the sake of simplification—I take full responsibility for any dogma.

Each chapter begins with a list of behavioral objectives and keywords. The keywords appear in bold type and are defined in the text. Chapters are internally organized—introductory principles and examples are given before applications. Figures and tables are designed to illustrate and organize important concepts presented in the text. Each chapter concludes with review questions that may be used by the student for self-testing.

Thanks to Jonna Hughes (mom), my personal and also professional proofreader, whose capacity to use a red pencil is only exceeded by her kindness to family and friends. I am indebted to Kathryn Dowling, PhD, Jon Kindschy, REHS, REA, RHSP, and Dennis Woodland, PhD, for providing scientific reviews. Thank you to Dr. Joyce Hopp, Dean, and Dr. Edd Ashley,

Associate Dean, School of Allied Health Professions (LLU), for encouragement and resources; Jerry Daly, Director of Media Services (LLU), for facilities and resources, and Louise Ceccarelli, Supervisor of Computer Graphics (LLU), for artistic design and numerous illustrations. Completion of the manuscript was facilitated by the individual efforts of Jan Fisher, Shirley Graves, Stacey Hughes, Gunter Reiss, Betsy Pavlick, Becky Pendergrass, Derek Reid, Lynn Steil, and Alan Swarm—thank you!

At Taylor & Francis I thank Richard O'Grady, Acquisitions Editor, for believing in the worthiness of my goal, and for providing technical support—especially prompt Internet replies to my queries; Carolyn Ormes, Development Editor, for midcourse corrections in style and deadline reminders; Bonny Gaston, Manufacturing Manager; Catherine Simon and Sharon M. Twigg, Production Editors; and Elizabeth Dugger, Copy Editor.

Asa once remarked that, in his native New Zealand, a man was considered successful if he built a house, had a son, and wrote a book. I built the house; have three wonderful daughters (Stacey, Summer, and Courtney) and a lovely, supportive wife (Marilyn); and wrote the book . . . he never told me if two out of three counted! However, my goal *is* achieved if students who read this book gain an understanding of how toxicants in the environment have their ultimate impact on the health of all organisms in the ecosystem—including our own species.

Billy Hughes
Loma Linda University
bhughes@ccmail.llu.edu

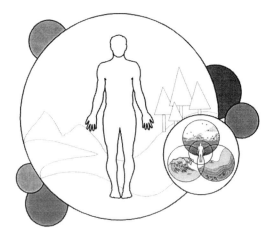

O bjectives

- Define environmental toxicology

- Describe the prehistory and history of toxicology

- Distinguish descriptive, mechanistic, and regulatory disciplines of toxicology

- Recognize the multidisciplinary approaches to environmental toxicology

- Summarize the relevance of environmental toxicology to the human species

K eywords

antidotes
atmosphere
biosphere
cells
clinical toxicology
descriptive toxicology
Ebers papyrus
ecosystem
environmental toxicology
etymology
forensic toxicology
hazardous waste
hydrosphere
industrial toxicology
infinite dilution
lithosphere
macromolecules
mechanistic toxicology
molecules
morbidity
mortality
Orfila
organ system
organelles
organs
Paracelsus

Keywords *(continued)*

phytotoxin

poisons

regulatory toxicology

tissue

toxic

toxicants

toxicity

toxins

venom

xenobiotics

What Is Environmental Toxicology?

Environmental toxicology is the study of the poisons around us. A general definition of environmental toxicology would include the hazardous effects that these poisons have on human health. Specifically, "environment" comes from the French word *environ*, which means "around," and *mens* is Latin for "mind." The word "toxicology" comes from the Greek word *toxikon*, a poisonous substance into which arrowheads were dipped, and the suffix *-logy* from the Greek word *logos*, which means the study of, or treatise.

Some toxicology terms appear to be similar but should be used with specificity to allow for accurate communication. For example, poisons are substances that in relatively small doses act to destroy life or seriously impair cellular function. There are a variety of poisons, many of which occur naturally in plants and animals or as minerals. There are also man-made poisons, which are the direct result of laboratory synthesis. Toxins are poisonous substances produced by plants (phytotoxins), animals (zootoxins), or bacteria (bacteriotoxins); a substance is toxic when it acts to destroy or impair cellular function. Toxicity is the state of being poisonous. The term venom refers to poisonous substances secreted by certain animals, such as bees, spiders, and snakes. When substances produce symptoms that are popularly referred to as intoxication (or poisoning) they are referred to as toxicants. There are naturally occurring toxicants, as well as toxicants that result from technological advances involving the manufacture and use of chemicals in industry and agriculture. Xenobiotics (Greek *xenos*, a stranger; *-biotic*, pertaining to life) may include substances, such as toxicants, that are not naturally produced within an organism.

Terminology

Learning the vocabulary of environmental toxicology allows for effective and accurate communication. Etymology is the study of word origins. The majority of environmental toxicology terms have their origins in classical Greek (G.) and Latin (L.). In addition to scientific terminology, these "dead" languages are also the source of many common words found in the English language.

Delving into a new discipline, such as environmental toxicology, requires that you learn the vocabulary along with the definitions. A knowledge of how words are formed can substantially reduce the amount of time spent in learning the meaning of new words. In general, scientific terms may contain three parts: a prefix, combining form, and suffix. To define a new term, start on the right and define the suffix, then move to the left, or beginning of the term, and define the prefix or combining form. For example, the word phytotoxin is made up of the combining form *phyto*, which comes from the G. *phyton*, a "plant," and *toxin* from the G. *toxikon*, which means "poison"—phytotoxin then is "poison from a plant."

A good scientific or medical dictionary is an invaluable aid when learning a new discipline. If you take the time to learn the parts of words you will be surprised at your ability to define new words without having to look them up in a dictionary.

History of Toxicology

The prehistoric use of animal venoms (L. *venenum*, venom) and plant poisons (L. *potio*, potion) is evident from archaeological and cultural anthropological studies. Ancient cultures had a working knowledge of many naturally occurring toxins that were used as medicinals, in hunting, and for war. Today, there are indigenous native peoples who still use naturally occurring poisons and toxins for hunting and for medicinal purposes.

One of the oldest written records of the early use of toxins is a series of eight Egyptian papyri dating from 1900–1200 B.C. The Ebers papyrus (Figure 1-1), which dates from 1500 B.C., contains directions for the collection, preparation, and administration of more than 800 medicinal and poisonous recipes. Some have obvious medicinal value, such as the use of opium to alleviate pain. The list also includes names of many potions of dubious medical value. Of interest to

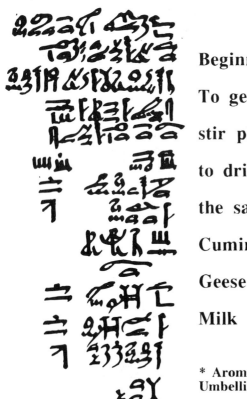

Beginning of the Book of Medications

To get rid of the diseases of the body

stir plant with vinegar

to drink by the patient

the same for the stomach, which is ill

Cuminum* (fraction of g)

Geese lard (fraction of g)

Milk (up to 0.6 l)

* Aromatic seeds from *Cuminum cyminum* (family Umbelliferae), a dwarf plant native to Egypt and Syria

Figure 1-1. A small portion of the Ebers papyrus. This Egyptian document, written about 1500 B.C., contains the recipes for more than 800 prescriptions. It also describes about 700 drugs of animal, vegetable, and mineral origin. (English translation courtesy of Dr. Gunter Reiss.)

toxicologists are recipes for poisons like hemlock, which was extracted from the dried unripe fruit of the plant *Conium maculatum* (Figure 1-2).

The use of plant and animal toxins by the Greeks was common. Dioscorides (A.D. 50–100), a Greek army physician who served in the court of Nero, the Roman emperor, is responsible for an early attempt to classify poisons. His classification of more than 600 plant, animal, and mineral poisons as being toxic or therapeutic is sufficiently valid to still be used today. The Greeks used

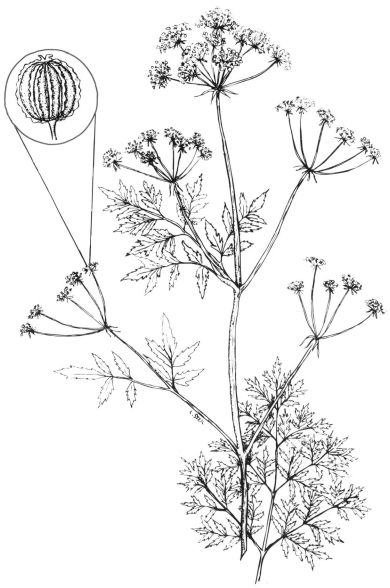

Figure 1-2. Poisonous hemlock can be extracted from the dried unripe fruit (circled) of *Conium maculatum*. (Original artwork by Lynn Steil.)

poisons as the state method of execution. Socrates, Demosthenes, and Cleopatra were all victims of poisoning, albeit for different reasons, including an execution and two suicides, respectively. Stories exist that the discovery of antidotes, which are agents to neutralize the effects of a poison, was facilitated by giving known toxins to condemned criminals followed by the administration of possible antidotes. When criminals survived these potentially lethal experiments, new antidotes could be added to the list.

The Romans (A.D. 50–400) made use of poisons for executions and assassinations, political or otherwise. By the fourth century A.D. the use of poisons had reached epidemic proportions. Did Agrippina kill Claudius to "clear the way" for Nero to become emperor of Rome? Did Nero subsequently kill Brittanicus, Claudius' natural son, with a soup *à la arsenic*?

The Islamic empire, following the death of Mohammed in A.D. 632, was strongly influenced by Greek medicine. Avicenna (A.D. 980–1036), a master of many disciplines, was considered to be an Islamic authority on poisons and their antidotes.

Ancient Chinese literature contains abundant references to the medicinal value of numerous plants and the poisonous properties of certain fish. Emperor Shen Nung (ca. 2700 B.C.) is reported to have experimented with poisonous as well as medicinal plants. Indeed, a "cure" for stupidity involved the use of poisons from newts and salamanders.

Hindu medicine in India from 800 B.C. to A.D. 1000 makes references to poisons and antidotes, such as for snake bites. Significant written works on medicine are found in the *Ayurveda*. Although the Hindus are believed to have borrowed some medicines from the Greeks, there are indications that the Greeks had medicinals of known Hindu origin.

Throughout the Middle Ages, poisons were used to gain political and social, as well as financial, advantage. In Italy, the Borgia family, including Cesare, his half-sister Lucretia, and their father Pope Alexander VI reportedly gained wealth as a result of their timely use of arsenic in wine. The term "lucre" (L. *lucrum*, gain), originally used to describe riches or wealth, is used now in a humorously derogatory sense. In A.D. 1198, Maimonides, a Spanish rabbi, wrote *Poisons and Their Antidotes*, which was a first-aid guide to the treatment of poisonings.

In general, the scholarship of the Middle Ages, from the ninth to fifteenth centuries, was based more on dogma than on empirical evidence. The German physician Paracelsus (1493–1541), a product of the Renaissance, brought the study of medicine and science to a new high (Figure 1-3). The role of experimentation, the relationship between dose and therapeutic, as compared with toxic, responses to chemicals, and the specificity with which different doses of chemical agents produce well-defined toxic or therapeutic effects are included in his writings—"What is there that is not a poison? All things are poison and nothing without poison. Solely the dose determines that a thing is not a poison." These early contributions form the basis of what is now the science of toxicology.

Of particular interest to environmental toxicology are the writings of the Italian physician Ramazzini. *Diseases of Workers*, published in 1713, deals with ailments that result from exposure to toxic chemicals in the workplace.

Figure 1-3. The German physician Paracelsus (1493–1541), whose writings promoted the essence of modern toxicology. (From *Paracelsus: Philosophia Magna.* Birckmann, Cologne, 1567.)

The beginnings of toxicology are usually traced to Orfila (1787–1853). With his 1815 book, *A General System of Toxicology, or, A Treatise on Poisons, Drawn from the Mineral, Vegetable, and Animal Kingdoms, Considered as to Their Relations with Physiology, Pathology, and Medical Jurisprudence*, this Spanish physician unknowingly established toxicology as a separate and distinct scientific discipline (Figure 1-4).

Subdisciplines of Toxicology

Each day toxicology impacts your life. A trip to your family physician may result in the use of chemical agents to aid in diagnosis, such as contrast agents used to enhance radiographic images. Or you may be prescribed a pharmaceutical agent that will prevent or treat disease. Excessive amounts of these substances can pose a danger to your health. The vegetables you eat may contain chemicals, both those used initially to promote pollination or growth and those added later to prolong shelf life. Animals slaughtered to provide meat may have been treated with chemicals to promote growth. If you check under the kitchen sink or in the garage you probably will find pesticides used to kill insects in the garden or molds growing on the shower tile. Each breath of air you take or glass of water you drink can potentially contain toxicants from a variety of industrial, automotive, agricultural, household, or natural sources.

A

GENERAL SYSTEM

OF

TOXICOLOGY,

OR,

A TREATISE ON POISONS,

DRAWN FROM

THE MINERAL, VEGETABLE, AND ANIMAL KINGDOMS,

CONSIDERED AS TO THEIR RELATIONS

WITH

PHYSIOLOGY, PATHOLOGY, AND MEDICAL JURISPRUDENCE.

By M. P. ORFILA, M. D.

OF THE FACULTY OF PARIS, PROFESSOR OF CHEMISTRY AND NATURAL
PHILOSOPHY.

TRANSLATED FROM THE FRENCH.

Unicum signum *certum* dati veneni est *notitia botanica* inventi veneni vege-
tabilis, et *criterium chemicum* dati veneni mineralis.

PLENCK. *Toxicologia.*

IN TWO PARTS.

London:

PRINTED FOR E. COX AND SON, ST. THOMAS'S-STREET,
BOROUGH.

1816.

Figure 1-4. Title page and a selected Contents page from an 1816 translation of the Spanish physician Orfila's 1815 book. (Courtesy of Library of the College of Physicians of Philadelphia.)

viii CONTENTS.

Figure 1-4. *(continued)*

Modern toxicology is composed of three subdisciplines. The first, descriptive toxicology, involves toxicity testing of chemicals. Initially, the determination as to whether or not a chemical is toxic must be made before safety and regulatory issues can be addressed. Toxicity testing usually takes place using experimental animals. Second, mechanistic toxicology examines the biochemical processes by which identified toxicants have an impact on the organism. Although descriptive toxicologists continue to identify agents of toxicity, the exact mechanism by which many toxicants have their action on the organism awaits continued study. Last, regulatory toxicology is concerned with assessing the data from descriptive toxicology and mechanistic toxicology in an attempt to determine the legal uses of specific chemicals, as well as the risk posed to the ecosystem by the marketing of those chemicals.

Many disciplines contribute to an understanding of toxicology (Table 1-1). Of particular interest, clinical toxicology examines the effects of toxicants on individuals and the efficacy of treatment for symptoms related to intoxication. Forensic toxicology is concerned with the medical and legal questions relating to the harmful effects of known or suspected toxicants, and industrial (or occupational) toxicology studies the disorders found in individuals who have been exposed to harmful materials in their place of work.

The scope of this book is environmental toxicology, which deals with the impact of known or suspected toxicants on the ecosystem, including the human population . . . the health hazards posed by the poisons around us. Although this text focuses on the effects of toxicants on the human species, remember that our ecosystem is complex, and potentially all forms of life, both plant and animal, may be affected by toxic substances.

Ecological Concepts

The biosphere is that region of our planet that contains living organisms (Figure 1-5). Although planet Earth is

Table 1-1. Selected disciplines that contribute to a more complete understanding of toxicology

Ecotoxicology	Industrial toxicology
Forensic toxicology	Risk assessment
Genetic toxicology	Environmental toxicology
Teratology	Biochemical toxicology
Cancer research	Immunotoxicology
Pesticide toxicology	Food toxicology
Analytical toxicology	Clinical toxicology
Veterinary toxicology	Vasculotoxicology
Hepatotoxicology	Pulmonotoxicology
Pharmacology	Industrial hygiene
Biochemistry	Epidemiology
Pathology	Medicine
Environmental law	Atmospheric sciences
Marine biology	Ecology
Soil sciences	Genetics
Cytology	Molecular biology
Cellular physiology	Physiology

Figure 1-5. The biosphere includes those regions of the atmosphere, hydrosphere, and lithosphere where living organisms are found.

quite large, the region where life can be found is just a thin veneer on the earth's surface, usually involving only a few meters of the lithosphere and a few kilometers of the atmosphere and hydrosphere. The ecosystem is a self-regulating community of animals and plants interacting with one another (biotic interactions) and with their non-living environment (abiotic interactions). Our ecosystem does not exist in a vacuum. Every activity, process, and interaction influences the ecosystem—much like the ripples that result when a stone is thrown into a pond go out and disturb the pond. When biotic or abiotic factors in the ecosystem are disturbed, they in turn will influence the interacting populations to either grow or decline in their numbers. These are often referred to as positive or negative feedback mechanisms, respectively.

Since time is a fundamental variable to everything that happens in an ecosystem, the time-dependent rates for changes to biotic and abiotic interactions in the ecosystem are important questions to consider—how fast the "ripples" travel. Our concept of time affects the questions we ask and the solutions we propose. A *very short time* may involve the instantaneous changes that last less than seconds, such as when lightning discharges or chemical reactions take place. *Human time* is usually the span of time in which most of us think, and it is in this unit of time that we make most of our observations about the ecosystem. "Long-term" planning within the constraints of the human time frame rarely exceeds 20 years. *Historical time* involves intervals of time that are too long to be studied by individuals. It is dependent on records provided by earlier generations. *Geological time* refers to long-term changes within the ecosystem. Within this time frame the processes that shape our earth and influence the structure of the ecosystem are measured. The drifting of continents, mountain building, and geological cycles of erosion and deposition of sediments are all processes that require long periods of time. Then there is *deep time* or stellar time—the time in which the universe exists, stars begin and end, and planets are formed.

Not surprisingly, our "long-term" planning is at odds with the stability of our ecosystem. The changes we precipitate are occurring at rates that are too fast, too overwhelming, for an ecosystem shaped over millions of years. The fact that environmental toxicology has developed into a significant discipline indicates that we have exceeded the self-regulatory ability of our ecosystem. To evaluate the profound impact we have made on the environment, including the relevance of environmental toxicology to our survival as a species, requires an understanding of (1) the sources of toxicants, (2) environmental cycles that transport toxicants, (3) the modes by which these toxicants enter and affect the human body, and ultimately (4) the degree to which society defines safety and what risks are acceptable as related to our exposure to toxicants.

Relevance of Environmental Toxicology to the Human Species

The human species has had a significant impact on the ecosystem. Apparently, early humans were better able to coexist with other animal and plant populations, most likely due to their small population size and reduced demands on the ecosystem. However, with the recent rapid increase in the size of the human population (including advances in industry and transportation, and economies based on continued growth) our once benign interaction with the ecosystem has changed into one where the demand for resources, such as food, water, and habitable space, is exceeding the supply.

An end product of the consumption of these resources is a tremendous amount of waste. For many decades infinite dilution was a common solution to the problem of waste disposal. Vast oceans (hydrosphere), land (lithosphere), and air (atmosphere) (Figure 1-5) were the "buckets" in which "drops" of potentially toxic wastes were diluted—indeed, waste disposal was viewed as just that—a "drop in the bucket." It was further thought that the toxic wastes would be of no consequence or harm to human health since the "buckets" were so vast. Somehow we forgot that eventually we would be exposed to the cumulative effect of the "drops" of toxic substances as we eat, breathe, and sleep in the "buckets."

Hazardous waste is defined as waste that, because of its biological, chemical, or physical characteristics, or quantity or concentration, may pose a danger of morbidity (disease) and mortality (death) to organisms (Figure 1-6). To illustrate the magnitude of the problem, in the United States alone over 4 billion tons of waste is generated each year from mining, agriculture, industry, and city sewage sources. On an average, each person contributes over 4 pounds of domestic solid waste each day. Of the approximate 274 million metric tons (1 metric ton = 2,200 lb) of this waste, which is subject to regulation as hazardous waste by the Environmental Protection Agency (EPA), only an estimated 10% is disposed of in an environmentally safe manner.

With over 5 million natural and man-made chemicals and over 80,000 synthetic chemicals currently being used in industry, agriculture, household, and other applications, the potential for exposure to hazardous waste is a concern. The health hazards for individuals who are exposed to hazardous wastes, as well as other toxic substances, when they are disposed of in an unsafe manner, poses a serious problem. This is especially true when cause-and-effect relationships are established between certain wastes and diseases. For example, based on worldwide epidemiological data the World Health Organization (WHO) estimates that 90–95% of all cancers are "environmen-tally related"—an environment we have disturbed is now afflicting our own population with morbidity and mortality.

Although contact with toxic substances may come as a result of occupational, accidental, or intentional exposure, there are some contacts or exposures over which we have little or no control. The saying that you can run but you can't hide is certainly true for aspects of the atmosphere, hydrosphere, and lithosphere on which we are dependent for our survival. We may unknowingly breathe air, drink water, or eat food that was polluted with toxicants hundreds or thousands of miles away.

The data of environmental toxicology should prompt us to stop or limit the sources of those substances that threaten to harm plant and animal species in the ecosystem. As a result of descriptive, mechanistic, and regulatory toxicological study, we will be better able to provide for the future of our species. It is with an awareness of these factors that environmental toxicology has its relevance to the human population.

Structural Levels of Organization

Nature can be organized into different levels of structural complexity, from subatomic particles to the ecosphere (Figure 1-7). Carbon (C), hydrogen (H), oxygen (O), nitrogen (N), calcium (Ca), potassium (K), and sodium (Na) are a few of the atoms essential to living organisms. When two or more atoms join together, molecules are formed. The molecules may be small (e.g., amino acids, simple sugars) or they may combine to form larger molecules called macromolecules, such as proteins. Cells are composed of complex

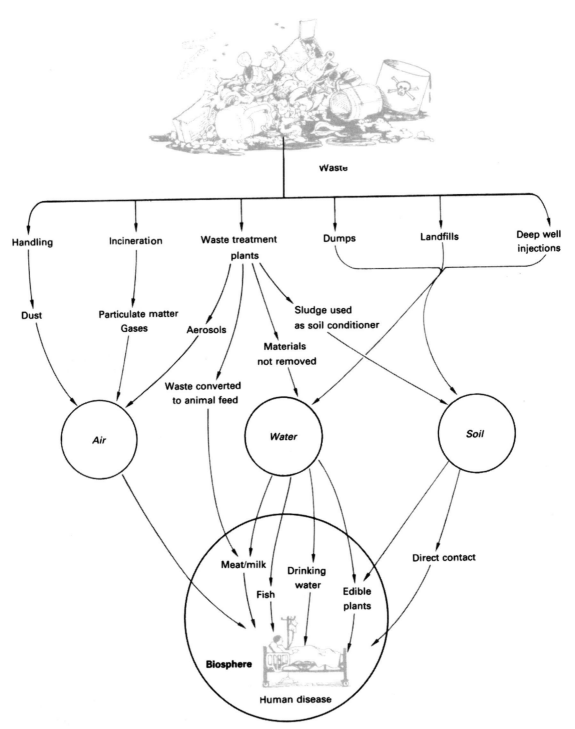

Figure 1-6. From waste to disease. The paths by which wastes, including toxicants, move into the biosphere where they produce morbidity. (From C. E. Kupchella and M. C. Hyland, *Environmental Science*. Allyn and Bacon, 1989. Reprinted by permission.)

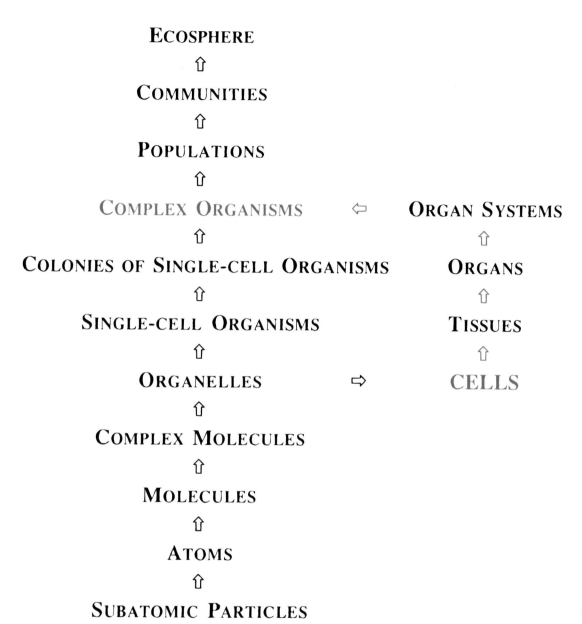

Figure 1-7. Levels of structural and functional organization.

assemblages of atoms, molecules, and complex molecules.

Cells are the basic unit of structure and function in a living organism. All of the functional or physiological processes in organisms ultimately take place at the level of the cell. Within the cell are small

structures called organelles that carry on specific activities.

Cells may be organized into units that together perform a similar function. These assemblages of cells are termed a tissue. There are four distinct tissue types in the human body: epithelial tissue, which covers the body and lines ducts and vessels; connective (or support) tissue, such as blood, bone, and collagen; muscle tissue, with smooth, skeletal, and cardiac types; and nerve tissue, which includes neurons.

Organs result when groups of different tissues unite to form structures that perform a specific function. Two or more organs may combine to form an organ system. For example, the use of food resources by the human body is accomplished by numerous organs (the gastrointestinal system), each of which functions in sequence to permit ingestion (e.g., oral cavity), mechanical and chemical digestion (e.g., stomach), absorption of nutrients (e.g., intestines), and elimination (e.g., anus).

Toxicants produce toxic effects by interacting with the molecules on or near the surface of, or within, the cell. The interactions in turn cause reversible or irreversible cellular damage by affecting proteins associated with the cell membrane (e.g., receptors), interfering with a cell's energy production (e.g., metabolism), binding to molecules within the cell (e.g., enzymes), or causing certain cells to die. Although environmental toxicologists may observe the gross effects of a toxicant on the whole organism, remember that the cumulative effect of the *disruption of structure and function at the level of the cell* is ultimately responsible for organismal morbidity and mortality.

Review Questions

1.   A general definition of this term would include a study of the hazardous effects that the poisons around us have on human health:

A.   Descriptive toxicology
B.   Environmental toxicology
C.   Forensic toxicology
D.   Mechanistic toxicology
E.   Regulatory toxicology

2.   Which term has its origin in the Greek word for "stranger"?

A.   Poison
B.   Toxin
C.   Toxicant
D.   Venom
E.   Xenobiotic

3.   This Greek physician served in the Roman emperor Nero's court and is responsible for classifying more than 600 plant, animal, and mineral poisons as being toxic or therapeutic.

A.   Avicenna
B.   Dioscorides

C. Maimonides
D. Ramazzini
E. Shen Nung

4. Identify the source of this statement: "All things are poison and nothing without poison. Solely the dose determines that a thing is not a poison."

A. The *Ayurveda*
B. The Ebers papyrus
C. Maimonides
D. Orfila
E. Paracelsus

5. Which area of toxicology is concerned with assessing the risk involved in the marketing of chemicals and their legal uses?

A. Descriptive toxicology
B. Forensic toxicology
C. Industrial toxicology
D. Mechanistic toxicology
E. Regulatory toxicology

6. "Long-term" planning within the constraints of the human time frame rarely exceeds 20 years.

A. True
B. False

7. It is estimated that 10% of hazardous wastes are disposed of in an environmentally safe manner.

A. True
B. False

8. In the human body, physiological processes ultimately take place at which level?

A. Cell
B. Tissue
C. Organ
D. Organ system
E. Organism

9. List five disciplines that contribute to a more complete understanding of toxicology.

10. Construct a diagram that shows the relationship between hazardous wastes, and morbidity and mortality.

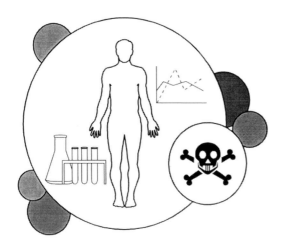

Objectives

- Define toxicity

- Discuss the different types of toxicity

- Describe toxicokinetics and toxicodynamics

- Explain how toxicants are classified

- Outline the steps involved in toxicity testing

Keywords

absorption
acute toxicity
biotransformation
chronic toxicity
delayed toxicity
distribution
elimination
end effect
immediate toxicity
in vitro
in vivo
linear dose sequence
local toxicity
logarithmic dose sequence
storage
systemic toxicity
toxicity testing
toxicodynamics
toxicokinetics

Toxicity

Toxicity, the state of being poisonous, is also a general term used to indicate adverse effects or symptoms produced by poisons or toxicants in organisms. Toxicity will vary according to both the duration and location of exposure to the toxicant, as well as the species-specific responses of the organism. Four distinct types of toxicity characterize the duration and location of the poisonous state. Acute toxicity involves a sudden onset of symptoms that last for a short period of time, usually less than 24 hours. The cellular damage that produces the symptoms associated with acute toxicity is usually reversible, resulting in recovery by the organism from the adverse effects brought on by the toxicant. Chronic toxicity results in symptoms that are of a long, continuous duration. The permanent nature of chronic toxicity is due to the irreversible cellular changes that have occurred in the organism. If cellular destruction and the related loss of function are severe, the organism may die.

Local toxicity occurs when the symptoms are restricted to the site of initial exposure to the toxicant. However, when the adverse effects occur at sites far removed from the initial site of exposure the term systemic toxicity is used. The ability for toxicants to be absorbed at one site and distributed to a distant region, such as an organ, results from transportation within the organism via the blood or lymphatic circulatory systems.

Exposure to carbon tetrachloride (CCl_4), an organic solvent used in industry, provides an example of the different types of toxicity. At high concentration for a short period of time, exposure to carbon tetrachloride vapors may result in toxicity and could involve minor eye and throat irritations. Upon cessation of this short-term, acute exposure to the vapors, the symptoms associated with this local toxicity will stop. However, if the exposure to carbon tetrachloride again involves a short period of time, but the toxicant is now absorbed through the skin or oral and ocular mucosa, it may enter the bloodstream and be transported via the blood to the brain. Once in the brain the toxicant produces symptoms, such as depression of the central nervous system (CNS), that may result in loss of consciousness. If the duration of exposure is short or acute, then cellular damage is reversible. Repeated exposure to high concentrations of either the vapor or liquid forms of carbon tetrachloride is capable of producing chronic and systemic toxicity, which is irreversible. Pathologies associated with chronic toxicity include kidney and liver damage, as well as severe CNS depression, which can lead to death.

Toxicity is also classified according to the timing between exposure to the toxicant and the first appearance of symptoms associated with toxicity. Immediate toxicity results when the symptoms occur rapidly within seconds or minutes following exposure to the toxicant. With immediate toxicity, the relationship between causative agents or toxicants and the pathologic symptoms or toxicity is more easily established. However, some toxicants may take years to produce toxicity. This delayed toxicity adds to the difficulty in establishing the cause-and-effect relationship. For example, diethylstilbestrol (DES) is a nonsteroidal drug prescribed for women during pregnancy to prevent miscarriage. It is now known that daughters born to mothers who took DES are at risk for developing vaginal and cervical cancers during adolescence. In this example the timing between *in utero* exposure to the toxicant and the first appearance of symptoms associated with toxicity may exceed 10 years.

It is important to recognize that toxicity results from exposure to specific toxicants and that the terms *acute* and *chronic* may also be used to describe the duration of exposure (e.g., acute exposure and chronic exposure). Evidence shows that acute and chronic exposure to many toxicants will parallel acute and chronic toxicity; however, it should be emphasized that in some cases acute exposure can lead to chronic toxicity. Finally, the terms that describe the type of toxicity may be used in combination, depending on the duration and location of toxicity and the timing between toxicant exposure and toxicity. An understanding of toxicity (i.e., acute, local, and immediate as compared to chronic, systemic, and delayed) should assist in characterizing the effects that toxicants have on organisms.

Toxicokinetics and Toxicodynamics

Toxicokinetics is the study of five time-dependent processes related to toxicants as they interact with living organisms. These processes are: absorption, how toxicants enter the organism; distribution, how toxicants travel within the organism; storage, how some tissues preferentially harbor a toxicant; biotransformation, how toxicants are altered (or detoxified) by chemical changes in the organism; and elimination, how toxicants are removed from the organism (Figure 2-1). An understanding of the time-dependent behavior of a toxicant as related to its absorption, distribution, storage, biotransformation, and elimination is necessary to explain how toxicants are capable of producing local or systemic toxicity, acute or chronic toxicity, and immediate or delayed toxicity.

Toxicodynamics examines the mechanisms by which toxicants produce unique cellular effects within the organism (Figure 2-1). As expected, if toxicants exert their influence at the level of the cell, the mechanisms will involve cellular components. Included in the mechanisms of toxic action are alterations to the cell's plasma membrane, organelles, nucleus, cytoplasm, enzyme systems, biosynthetic pathways, development, or reproduction. Whether reversible or irreversible cellular injury occurs will depend on the duration of exposure as well as the specific toxicokinetic properties of the toxicant.

Classification of Toxicants

Many classification schemes for toxic agents have been proposed (Table 2-1). Dioscorides classified substances using the general characteristics of whether they were toxic or therapeutic. Additionally, the source of the toxicant has long been recognized as a means of classification. An early scheme by Orfila classified substances as being of animal, vegetable, or mineral origin. No single classification system can be expected to adequately distinguish all known toxicants. As more data related to toxicants becomes available there will

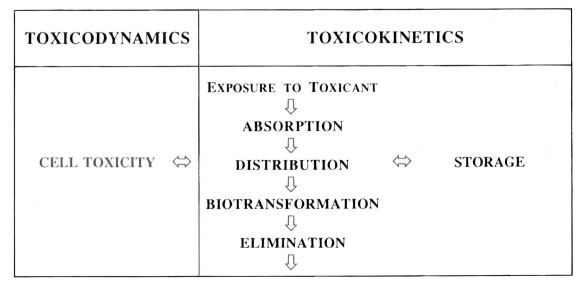

Figure 2-1. Simplified schematic representation of the toxicokinetic and toxicodynamic processes that connect exposure to a toxicant to the resulting toxicity.

undoubtedly be more characteristics that can be used for classification purposes. It is important to pay careful attention in the selection and presentation of a classification system or combination of systems so they will be informative and appropriate for the intended audience.

For example, a highly technical mechanism of action classification may be meaningless to a general nonscientific audience. Ultimately, the value of any classification system is its ability to adequately convey comparative, relative, or absolute information.

Table 2-1. Some of the ways commonly used
to classify toxicants

Classification	Categories
Physical state	Gas, liquid, solid, dust
Use	Pesticide, solvent, food additive
Chemical structure	Aromatic amines, aliphatics, glycols
General action	Air pollutants, chronic poisons, industrial toxins
Effect	Carcinogens, mutagens, teratogens
Target organ	Neurotoxins, hepatotoxins, nephrotoxins
Mechanism of action	Stimulants, inhibitors, blockers
Poisoning potential	Slightly toxic, moderately toxic, supertoxic
Labeling requirement	Oxidizer, acid, explosive
General or use class	Plastics, organic chemicals, heavy metals

Determination of Toxicity

Since toxicity is the state of being poisonous, it is imperative that cause-and-effect relationships between substances (in this case, suspected toxicants and the resultant toxicity) be established. The descriptive toxicologist's role is to identify and establish the cause–effect or, more specifically, the toxicant–toxicity relationship. Toxicity then is determined when, on the administration of a substance, an observable and well-defined end effect is identified. Paracelsus recognized the value of cause-and-effect relationships and the specificity with which different doses of chemical agents produced well-defined toxic or therapeutic effects.

Toxicity testing involves four steps. First, a *test organism* must be selected. Plants or animals can be used. Algae, bacteria, mice, rats, rabbits, or nonhuman primates are often selected as the test organisms. *In vivo* (in life) studies use the whole organism for toxicity testing. Humans, for moral and ethical reasons which are culturally defined, are normally not chosen as the test organisms. Current *in vitro* (in glass or test tube) studies do not use the whole organism but instead make use of cultured cells or tissue cultures, providing an attractive alternative in terms of cost and ethics to *in vivo* toxicity testing. It is important to consider the applicability of the test organism as related to generalizations that will undoubtedly result from the toxicity test; that is, will the conclusions based on a plant, bacterium, or rat provide the relevant information as related to humans?

Second, the *response* (end effect) to be observed and recorded must be selected. The response needs to be easily observable and quantifiable. Of the many possible responses, some that are commonly used include changes in the total number of cells in a bacterial colony, the presence or absence of biochemical products produced by cultured cells, changes in cell morphology, number of tumors produced, alterations in sleep patterns, and changes in growth and development of an organism (Figure 2-2). For *in vivo* studies, the death of the experimental organism is the end effect.

Third, a selection of the *duration of the test* or exposure period is necessary (Table 2-2). The duration may range from a few seconds to years, depending on the type of test being performed. Eye irritant tests may only take a few seconds, whereas reproductive studies may take years, particularly when multiple generations are examined.

Fourth, *doses* to be tested are selected. For *in vivo* studies the dose is expressed as the weight in milligrams (mg) of the substance being tested per kilogram (kg) of body weight of the experimental organism. This is written as mg/kg. Although a statement of absolute amounts of the substance being tested may appear to be a better way of quantifying the dose, it is not, since toxicity is related to the size of the organism. For example, 100 mg in a 250-g rat is very different from 100 mg in a 5,000-g monkey (Table 2-3). For *in vitro* toxicity testing, the weight in milligrams (mg) of the substance being tested per milliliter (mL) of medium containing the cells expresses the dose, written as mg/mL.

Once the test organism, responses to monitor, exposure period, and series of doses to test have been selected, toxicity

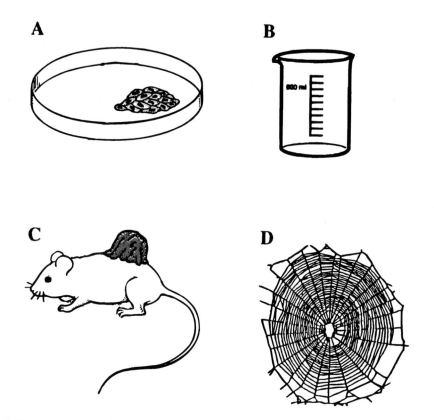

Figure 2-2. Examples of responses observed and measured during toxicity testing: (A) growth and differentiation of cells in tissue culture, (B) quantitative and qualitative changes in urine and blood, (C) formation of tumors, and (D) construction of webs by spiders.

Table 2-2. Selected descriptive toxicity tests as required for U.S. registration and estimated costs

Toxicity test	Estimated cost[a]
Acute eye irritation test (rabbit)	$ 500.00
Acute skin sensitization test (guinea pig)	1,000.00
Acute dermal toxicity test (rabbit)	5,000.00
Acute inhalation toxicity test (rat)	10,000.00
Repeated dose test: 14-day exposure (rat)	50,000.00
Repeated dose test: 90-day exposure (rat)	120,000.00
Repeated dose test: 1-year exposure (rat with dietary intake)	250,000.00
Repeated dose test: 1-year exposure (rat with force feeding)	1,000,000.00
Bacterial genetics test (reverse mutation)	3,000.00
Chromosome aberration test (rat)	30,000.00
Reproductive toxicity test (rat)	100,000.00

[a]Costs will vary depending on replications, duration of test, etc.

Table 2-3. A comparison of weight, dosage, and dose for selected animals used for *in vivo* toxicity testing (454 g = 1 lb)

Organism	Weight (g)	Dosage (mg/kg)	Dose (mg/animal)
Mouse	25	100	2.5
Rat	250	100	25
Guinea pig	500	100	50
Rabbit	1,500	100	150
Cat	2,500	100	250
Monkey	5,000	100	500
Dog	10,000	100	1,000
Human	75,000	100	7,500

testing can proceed. Only after the selection of these parameters and subsequent experimentation can the relationship between dose and response be established and the genius of Paracelsus' statement be appreciated: "*All substances are poisons; there is none which is not a poison. The right dose differentiates a poison and a remedy.*" Everything—water, oxygen, and chocolate included—may be considered a toxicant if the dose is great enough.

Determining the Doses to Test

The ability to accurately determine which doses are responsible for producing a specific end result is critical. If the selected doses *all* produced the predetermined response, then questions relating to the minimum dose required to produce the selected response cannot be answered. This is of particular concern when toxicity testing is done to determine the minimum dose required to produce a response. For this reason the series of doses selected usually will be in a logarithmic dose sequence instead of a linear dose sequence. For example, instead of testing linear doses of 1, 2, 3, 4, 5, 6, 7, 8, 9, and 10 mg/kg,

the logarithmic dose sequence of 0.01, 0.1, 1.0, 10, 100, and 1,000 mg/kg will be tested. Logarithmic doses have an advantage over linear doses in their ability to maximize the range of doses being tested, while minimizing the possibility of overlooking a small dose that may represent the response threshold or minimum dose at which the end effect will first be observed. Estimates of the range of doses to test for a suspected toxicant are often made from previous experience, including toxicity test results from similar chemical substances, or a range-finding subchronic study.

Variables Affecting Toxicity

A toxicant's ability to interact with cellular structures to produce morbidity and mortality depends on both the toxicant being tested and the chosen experimental organism. Characteristics of the toxicant include the intrinsic factors of the chemistry of the toxicant itself, as well as the concentration of the toxicant. The chemistry of the toxicant is usually well defined and limited in a chemical sense—that is, the toxicant will behave in a predictable manner.

In contrast to the toxicant's predictable chemical behavior, the organism's response to the toxicant will depend on a number of variables related to the rates of absorption, distribution, storage, biotransformation, and elimination. These variables are different for each species. Species specificity is evidenced in the effects of thalidomide, a sedative drug used in humans in the late 1950s to treat morning sickness associated with pregnancy, to produce the same toxicity in different organisms. This drug is capable of crossing the placental barrier to produce congenital malformations of fetal limbs, such as amelia (absence of limbs) and phocomelia (presence of seal-like flippers). However, not all mammalian species are equally affected by the drug, as seen in mice and rats, which are resistant, and hamsters and rabbits, which show variable effects.

In addition to species differences, there may be *gender* differences, with males and females exhibiting different responses to a toxicant. *Age* plays a role, as evidenced by different responses to toxicants by young, middle-aged, and older individuals. *Nutritional status*, particularly a lack of essential vitamins and minerals, can lead to impaired cellular function and render cells vulnerable to toxicants. *Disease states* may also affect the organism's interaction with toxicants. Finally, *time of day* of exposure may be important, as hormones and enzyme levels are known to fluctuate during the course of a day (i.e., circadian rhythms).

Review Questions

1. Which is *not* a correct statement about acute toxicity?

A. Involves a sudden onset of symptoms.
B. Symptoms typically last less than 24 hours.
C. Results in cellular damage that is irreversible.
D. The organism usually recovers from the adverse effects.
E. May involve local or systemic toxic responses.

2. With this form of toxicity it is often difficult to establish a cause-and-effect relationship:

A. Acute toxicity
B. Delayed toxicity
C. Immediate toxicity
D. Local toxicity
E. Systemic toxicity

3. Toxicokinetics is the study of all but which one of the following?

A. Absorption
B. Biotransformation

C.   Distribution
D.   Elimination
E.   Mechanism of toxicity

4.   Of the many ways commonly used to classify toxicants, the categories of stimulants, inhibitors, and blockers best characterize which classification scheme?

A.   Chemistry
B.   General
C.   Poisoning potential
D.   Mechanism of action
E.   Target organ

5.   Which one of the following represents a correct quantification of a dose as used in an *in vivo* toxicity test?

A.   10 mg
B.   10 mg/animal
C.   10 mg/kg
D.   10 mL
E.   10 mg/mL

6.   Which represents a logarithmic dose sequence?

A.   1, 2, 3, 4, 5, 6, 7 mg/kg
B.   1, 5, 10, 15, 20, 25 mg/kg
C.   10, 20, 30, 40, 50 mg/kg
D.   0.1, 1, 10, 100, 1,000 mg/kg
E.   0.1, 0.2, 0.3, 0.4, 0.5 mg/kg

7.   _____ examines the mechanism by which a toxicant has its unique cellular effect within the organism.

8.   Toxicity is determined when, on the administration of a substance, an observable and well-defined _____ is identified.

9.   Why is the correct selection of a test organism critical to a toxicity test?

10.   Discuss variables (e.g., age, gender, disease states) affecting toxicity and toxicity testing.

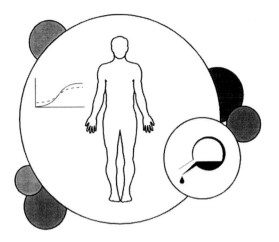

bjectives

- Explain the difference between causal and associative relationships

- Discuss the role of epidemiology in establishing associative relationships

- Describe the relationship between dose and response

- Interpret frequency and cumulative dose-response curves

- Recognize subthreshold, threshold, and ceiling effect doses

- Summarize effective, toxic, and lethal doses

- Define potency, efficacy, mixed or reversed toxicity, and margin of safety

eywords

associative relationships
causal relationships
ceiling effect
cumulative dose-response graph
dose-response relationship
ED_{50}
effective dose (ED)
efficacy
epidemiological studies
frequency dose-response graph
LD_{50}
lethal dose (LD)
margin of safety
mixed toxicity
No Observable Effect Level (NOEL)
normal distribution
potency
potent
reversed toxicity
subthreshold doses
TD_{50}
threshold dose
Threshold Limit Value (TLV)
toxic dose (TD)

Dose and Response Relationships

A dose-response relationship exists when a consistent mathematical relationship describes the proportion of test organisms responding to a specific dose for a given exposure period. Although a dose-response relationship may seem easy to establish, a number of assumptions need consideration.

The first assumption is that the observed response is caused by the substance administered. A causal relationship must be established between the dose administered and the observed response. It should be remembered that causal relationships are very different from associative relationships. Direct cause-and-effect linkages involving single variables, rather than associative linkages with two or more variables, are needed to establish the dose-response relationship. Retrospective epidemiological studies (i.e., prevalence of disease and death in a population) often conclude that an "associative" relationship exists between two observations. This may prompt efforts to establish an exact causal relationship, but a dose-response relationship cannot be concluded until a definitive causal relationship is demonstrated. Medical records (e.g., death certificates) are invaluable aids in establishing linkages between specific diseases or causes of death and their associative relationship to numerous variables (Figure 3-1).

Second, the magnitude of the response is assumed to be directly related to the magnitude of the dose. This assumption goes beyond the first assumption that the observed response is caused by the substance administered.

In this assumption, a mechanism of action is proposed that involves the cell and the myriad of molecules with which the substance being tested can interact to produce the observed response. It is assumed that there is a relationship between the dose administered and the eventual concentration of the substance as it interacts at the level of the cell. A descriptive toxicologist can usually determine that the first assumption is correct for a given toxicant; however, the mechanistic toxicologist ascertains the correctness of this second and more difficult assumption.

The third dose-response relationship assumption states that it is possible to correctly observe and measure a response. The ability to define and observe pathologies associated with toxicity is dependent on the depth of understanding of cellular anatomy and physiology. It is impossible to select and subsequently measure responses for which related cellular structures and processes are unknown. This assumption stresses the value of the contributions from basic science research. Disciplines such as cytology, genetics, molecular biology, and cellular physiology establish the cellular structural and functional norms against which pathologies resulting from toxicity can be compared.

Dose-Response Graphs

The graphic presentation of dose and response data permits an environmental toxicologist to readily determine important dose-response relationships. Furthermore, the graphs enable different toxicants to be compared. In a dose-response graph the

CERTIFICATE OF DEATH

1 NAME OF DECEDENT—FIRST (GIVEN) NAME OF DECEDENT | 2 MIDDLE | 3 LAST (FAMILY)

4 DATE OF BIRTH MM/DD/CCYY DATE OF BIRTH | 5 AGE YRS | IF UNDER 1 YEAR MONTHS DAYS | IF UNDER 24 HOURS HOURS MINUTES | 6 SEX | 7 DATE OF DEATH MM/DD/CCYY DATE OF DEATH | 8 HOUR

9 STATE OF BIRTH STATE OF BIRTH | 10 SOCIAL SECURITY NO | 11 MILITARY SERVICE MILITARY SERVICE 19 ___ TO 19 ___ NONE | 12 MARITAL STATUS MARITAL STATUS | 13 EDUCATION —YEARS COMPLETED EDUCATION

14 RACE RACE | 15 HISPANIC—SPECIFY YES ___ NO | 16 USUAL EMPLOYER USUAL EMPLOYER

17 OCCUPATION OCCUPATION | 18 KIND OF BUSINESS KIND OF BUSINESS | 19 YEARS IN OCCUPATION YEARS IN OCCUPATION

20 RESIDENCE—STREET AND NUMBER OR LOCATION RESIDENCE

21 CITY CITY | 22 COUNTY COUNTY | 23 ZIP CODE | 24 YRS IN COUNTY YEARS IN COUNTY | 25 STATE OR FOREIGN COUNTRY STATE

26 NAME RELATIONSHIP | 27 MAILING ADDRESS (STREET AND NUMBER OR RURAL ROUTE NUMBER CITY OR TOWN STATE ZIP)

28 NAME OF SURVIVING SPOUSE—FIRST | 29 MIDDLE | 30 LAST (MAIDEN NAME)

31 NAME OF FATHER—FIRST NAME OF FATHER | 32 MIDDLE | 33 LAST | 34 BIRTH STATE BIRTH STATE

35 NAME OF MOTHER—FIRST NAME OF MOTHER | 36 MIDDLE | 37 LAST (MAIDEN) | 38 BIRTH STATE BIRTH STATE

39 DATE MM/DD/CCYY | 40 PLACE OF FINAL DISPOSITION

41 TYPE OF DISPOSITION(S) | 42 SIGNATURE OF EMBALMER ▶ | 43 LICENSE NO

44 NAME OF FUNERAL DIRECTOR | 45 LICENSE NO | 46 SIGNATURE OF LOCAL REGISTRAR ▶ | 47 DATE MM/DD/CCYY

101 PLACE OF DEATH PLACE OF DEATH | 102 IF HOSPITAL SPECIFY ONE IP ER/OP DOA | 103 FACILITY OTHER THAN HOSPITAL CONV HOSP RES OTHER | 104 COUNTY COUNTY

105 STREET ADDRESS—STREET AND NUMBER OR LOCATION STREET ADDRESS | 106 CITY CITY

107 DEATH WAS CAUSED BY (ENTER ONLY ONE CAUSE PER LINE FOR A B C AND D) | TIME INTERVAL BETWEEN ONSET AND DEATH | 108 DEATH REPORTED TO CORONER YES NO REFERRAL NUMBER

IMMEDIATE CAUSE (A) IMMEDIATE CAUSE OF DEATH

DUE TO (B) DUE TO | 109 BIOPSY PERFORMED YES NO

DUE TO (C) | 110 AUTOPSY PERFORMED YES NO

DUE TO (D) | 111 USED IN DETERMINING CAUSE YES NO

112 OTHER SIGNIFICANT CONDITIONS CONTRIBUTING TO DEATH BUT NOT RELATED TO CAUSE GIVEN IN 107
OTHER SIGNIFICANT CONDITIONS CONTRIBUTING TO DEATH

113 WAS OPERATION PERFORMED FOR ANY CONDITION IN ITEM 107 OR 112? IF YES LIST TYPE OF OPERATION AND DATE

114 I CERTIFY THAT TO THE BEST OF MY KNOWLEDGE DEATH OCCURRED AT THE HOUR DATE AND PLACE STATED FROM THE CAUSES STATED
DECEDENT ATTENDED SINCE MM/DD/CCYY | DECEDENT LAST SEEN ALIVE MM/DD/CCYY
| 115 SIGNATURE AND TITLE OF CERTIFIER ▶ | 116 LICENSE NO | 117 DATE MM/DD/CCYY
118 TYPE ATTENDING PHYSICIAN S NAME MAILING ADDRESS + ZIP

I CERTIFY THAT IN MY OPINION DEATH OCCURRED AT THE HOUR DATE AND PLACE STATED FROM THE CAUSES STATED
119 MANNER OF DEATH
NATURAL SUICIDE HOMICIDE
ACCIDENT PENDING INVESTIGATION COULD NOT BE DETERMINED

120 INJURY AT WORK YES NO | 121 INJURY DATE MM/DD/CCYY INJURY DATE | 122 HOUR | 123 PLACE OF INJURY PLACE OF INJURY

124 DESCRIBE HOW INJURY OCCURRED (EVENTS WHICH RESULTED IN INJURY)
DESCRIBE HOW INJURY OCCURRED

125 LOCATION (STREET AND NUMBER OR LOCATION AND CITY AND ZIP CODE)
LOCATION

126 SIGNATURE OF CORONER OR DEPUTY CORONER ▶ | 127 DATE MM/DD/CCYY | 128 TYPED NAME, TITLE OF CORONER OR DEPUTY CORONER

Figure 3-1. Death certificates provide an important source of information used in retrospective epidemiological studies.

horizontal axis (X axis or abscissa) always represents the dose in a logarithmic scale using mg/kg units (Figure 3-2A and B). The vertical axis (Y axis or ordinate) represents the *in vivo* or *in vitro* response. Proper labeling, which indicates the response being measured and the unit used for the series of doses being tested, is necessary for accurate communication of the dose-response relationship. To avoid errors associated with misreading fractional doses, decimal points should be preceded by a zero (e.g., 0.1 mg/kg, instead of .1 mg/kg).

The "response" axis in a dose-response graph may be presented as a frequency-response or a cumulative-response. A frequency dose-response graph plots the percentage of organisms responding to a given dose (Figure 3-2A). These graphs can usually be recognized by their "bell-shaped" appearance. The cumulative dose-response graph represents the cumulative sum of responses from lower to higher doses (Figure 3-2B). The frequency or percentage of organisms responding to the lowest dose is added to the percentage responding to the next dose, which is then added to the third dose, and so on. The line on these graphs appears "sigmoidal," a name that comes from the Greek letter *sigma*, "σ." Cumulative dose-response graphs are often seen in toxicity studies related to environmental toxicology.

Statistical Considerations of Dose-Response

Any data that results from measurements on a sample (statistic) has a sampling distribution. It is assumed that the data generated by dose-response experimentation will follow a Gaussian or normal distribution. When a normal distribution is present, the resulting frequency-response graph will appear bell-shaped, whereas the cumulative-response graph will appear sigmoidal. Although the lines on these two graphs appear to be different, in reality they are just different graphic presentations of the same data.

Responses observed in actual test organisms are assumed to be representative of the total or universal population of potential test organisms. The validity of this assumption is questionable when small numbers of test organisms are used. On the other hand, it is not cost-effective to perform dose-response experiments when large numbers of test organisms are used. Somewhere between too many and too few test organisms a decision must be made as to the minimum number of test organisms (N) needed to establish a statistically valid dose-response conclusion. To make this decision requires information about the response variability within the population of potential test organisms, the desired statistical *strength* of any conclusions that may result, and available resources (including cost of tests, time for testing, available personnel, and laboratory space). By paying careful attention to research design, toxicologists can avoid the erroneous conclusions that may result from the use of too few test organisms during toxicity testing.

When the response measurements are normally distributed it is observed that the greatest number (frequency) of test organisms will exhibit the response at a dose somewhere between the lowest and highest doses tested. This is

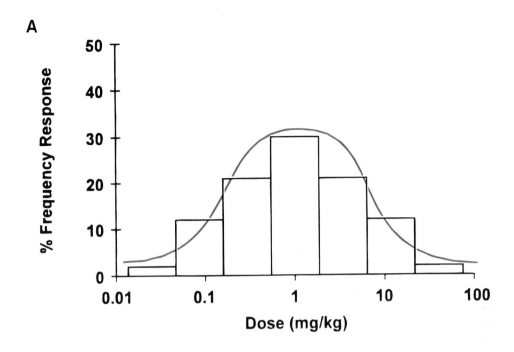

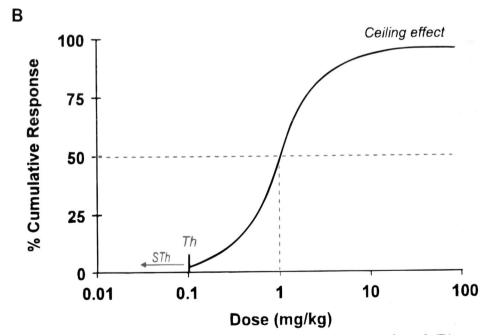

Figure 3-2. Dose-response graphs: (A) frequency-response graph and (B) cumulative-response graph showing subthreshold dose (STh), threshold dose (Th), and ceiling effect.

visually evident in the apex of the bell-shaped line on a frequency dose-response graph or in the middle flat region of the line on a cumulative dose-response graph. A specific point located in these regions represents the mean ($\overline{X}$) or average response and is equal to the sum of all responses ($\sum X$) divided by the number of responses (N), or $\overline{X} = \sum X/N$.

When the responses of test organisms follow a normal distribution there are always a few organisms in which the predetermined response will occur at a very low dose, as well as a few organisms that will not exhibit the response until a very high dose is given. These "supersensitive" or "hypersusceptible" and "hearty" or "resistant" test organisms are graphically represented on the sides of the bell and at the beginning and end of the sigmoidal line, respectively. The distance of these "outliers" from the mean (average) response is best stated by use of a statistic called the standard deviation (SD). A large or small SD value is able to convey valuable information about the dose-response relationship in a test population.

For starters, ±1 SD accounts for 67% of test organism responses, ±2 SD accounts for 95%, and ±3 SD represents 99% of test organism responses. Assume that the mean response is constant for two toxicants. If the SD for toxicant A is very large and the SD for toxicant B is very small, it can be concluded that there is a wide range of doses over which the test organisms responded to toxicant A as compared with the small range of doses over which the test organisms responded to toxicant B (Figure 3-3). Numbers such as the mean and standard deviation are

useful; however, changes in the shape of the bell-shaped line or in the sigmoidal line of the cumulative dose-response graph allow for rapid visual characterization and comparison of toxicants.

Three features characterize the sigmoidal line on a cumulative dose-response graph. First, there is a dose at which the first test organism will respond (Figure 3-2B). This is referred to as the threshold dose, which can be seen on the graph as the left-side beginning of the sigmoidal line. Subthreshold doses are represented to the left of this point. At these doses no responses were observed. The following are often used to refer to this beginning region of the cumulative dose-response graphs: No Observable Effects Level (NOEL), No Observable Adverse Effect Level (NOAEL), Suggested No Adverse Response Level (SNARL), Lowest Observable Effect Limit (LOEL), and Threshold Limit Value (TLV).

At progressively higher doses, the initially curved sigmoidal line begins to straighten out (Figure 3-2B). This second region of the graph represents the doses at which the majority of test organisms were observed to exhibit the response. Of interest is the cumulative 50% level, representing the dose at which the mean response occurred. Third, the right side of the line on a cumulative dose-response graph may be seen to once again curve and then become almost horizontal or flat (Figure 3-2B). This region represents the higher doses at which the remaining few test organisms finally exhibited the predetermined end effect. This region is said to exhibit the ceiling effect, since an increase in dose produces little or no increase in response.

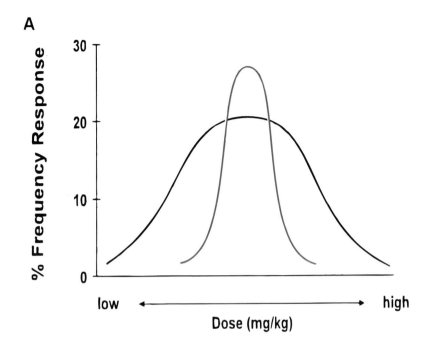

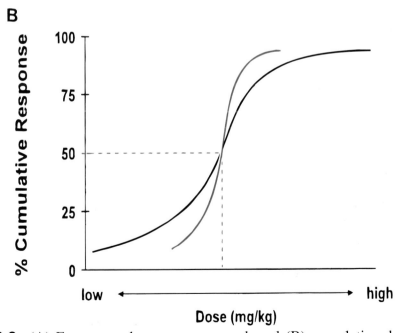

Figure 3-3. (A) Frequency dose-response graph and (B) cumulative dose-response graph showing data from two toxicity studies that have the same mean doses ($\overline{X}$) but different distributions (i.e., standard deviations).

The cumulative 100% level is indicated on the right side, at the point where the graph line stops.

Interpreting Dose-Response Data

Depending on the observable response selected for the toxicity test, dosages may be characterized as being effective, toxic, or lethal. An effective dose (ED) is evident when the desirable response is observed at the dose tested. A toxic dose (TD) represents the dose at which toxicity is present in test organisms. When lethality is the selected response, a lethal dose (LD) represents the dose resulting in the death of the test organism. LD is always used when lethality is selected as the observable response, even though lethality is in itself a response and could be referred to as ED. Note that *therapeutic dose*, although used in pharmacology, is considered an effective dose and should be signified by ED, not TD.

To facilitate the interpretation of data from a single dose-response study or when comparing data from two or more dose-response studies, it is useful to examine the dose at which a specified cumulative percentage of test organisms exhibit the ED, TD, or LD. A numerical subscript is added to denote the cumulative percentage of the test organisms that exhibit the predetermined response. For example, the ED_{50}, TD_{50}, or LD_{50} is indicative of the dose at which 50% of the test organisms were observed to exhibit the effective, toxic, or lethal response. Although 50% is often used for comparing toxicity, other cumulative percentages are also used such as ED_{99},

TD_{10}, or LD_{01}. Remember, valid comparisons between dose-response data from two or more toxicity studies require that the same cumulative percentage be used. In other words, don't compare the dose at which an LD_{01} occurs for toxicant A with the dose at which an LD_{50} occurs for toxicant B.

Determining relative toxicity is important. When comparing two toxicants, the one with the smaller ED_{50}, TD_{50}, or LD_{50} is considered to be the more potent (Figure 3-4A). This means that the observed response occurred at a lower dose, or a smaller dose of toxicant A as compared to toxicant B produced the same response. Potency then is a relative concept for comparing toxicants and, provided that the same response cumulative percentages are used (e.g., ED_{01}, ED_{05}, or ED_{99}), statements about potency are informative and may be used to classify toxicants. Of additional value is the term efficacy. A toxicant is said to have a higher efficacy when the dose-response relationship continues over a greater range of doses (Figure 3-4B). Some toxicants are capable of evoking a response in 100% of the test organisms over a short range of doses; however, other toxicants will continue to produce responses at even higher doses.

On occasion, the sigmoidal lines on cumulative dose-response graphs from two or more toxicity studies will be seen to intersect or cross, revealing a mixed or reversed toxicity relationship (Figure 3-4C). This occurs when one toxicant is not consistently more potent over the range of doses tested as compared to another toxicant—the dose-response curves cross. For example, the LD_{10} for toxicant A occurs at a higher dose than for toxicant B, but the LD_{50} for toxicant

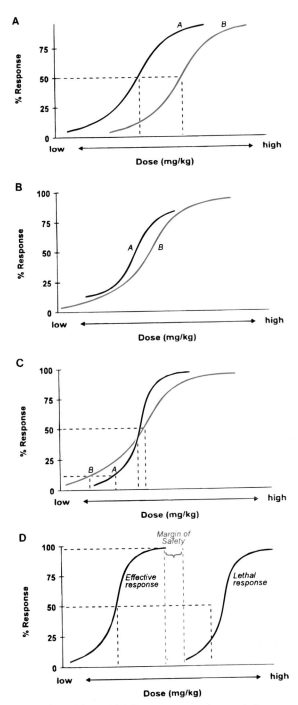

Figure 3-4. Toxicity relationships. (A) Potency: toxicant A is more potent than toxicant B; (B) efficacy: toxicant B is more efficant than toxicant A; (C) mixed or reversed toxicity: comparing the LD_{10}'s, toxicant B is more potent than toxicant A, but when comparing LD_{50}'s, toxicant A is more potent than toxicant B; and (D) margin of safety: determining the ratio of effective doses to lethal doses, for example, LD_{01}/ED_{99} or LD_{50}/ED_{50}.

A is lower than for toxicant B. Mixed toxicity relationships are not detected if only a single cumulative dose-response percentage is used for comparison (e.g., ED_{01}, TD_{10}, or LD_{50}). For this reason it is of value to examine cumulative dose-response data and graphs over the entire range of test doses.

Margin of safety expresses the magnitude of the range of doses between a noneffective or minimally effective dose (e.g., NOEL or LOEL) and a lethal dose (LD) (Figure 3-4D). The margin of safety is determined from the results of two toxicity studies, such as an ED study and an LD study. Dose-response data from the ED study is valuable as it will indicate the dose at which minor, acute, or reversible signs of toxicity are produced and thereby establish a threshold dose. Next, TD or LD dose-response studies are performed to establish the doses at which toxicity may be potentially chronic, irreversible, or lethal. A ratio between selected LD and ED values is used to express the margin of safety, such as LD_{01}/ED_{99} or TD_{50}/ED_{50}. The larger the ratio, the greater the margin of safety. These ratios are useful in determining what constitutes an acceptable environmental exposure. For pharmaceuticals, a large LD_{01}/ED_{99} ratio is also desirable, since it indicates that the therapeutic value of a drug can be obtained at relatively low doses as compared to the much higher doses at which the drug will be lethal. This is important, especially when the potential for overdose is a concern.

Review Questions

1. Type of relationship that exists when a constant mathematical relationship describes the proportion of test organisms responding to a specific dose for a given exposure period:

A. Associative
B. Dose-response
C. Epidemiological
D. Exposure
E. Toxicity

2. Which is not a true statement about epidemiology?

A. It is of value in determining associative linkages.
B. It involves study of the prevalence of disease and death in a population.
C. It may involve retrospective studies.
D. It is used to establish dose-response relationships.
E. It involves the study of two or more variables.

3. Identify the one false statement about cumulative dose-response graphs:

A. The graphs permit different toxicants to be compared.
B. The horizontal axis represents the dose.

C. The vertical axis represents the cumulative responses.
D. The graphs typically have a "bell-shaped" appearance.
E. When writing fractional doses the decimal should be preceded by a zero (i.e., 0.1).

4. When examining a frequency dose-response graph, where would you find "hardy" or "resistant" test organisms?

A. At doses represented by NOEL.
B. At the center of the bell-shaped curve at the mean dose.
C. At subthreshold doses.
D. At threshold doses.
E. On the extreme right side where the graph line stops.

5. On comparing two chemicals it is noticed that chemical A has a smaller LD_{50} than chemical B. This means that:

A. A has a greater efficacy than B.
B. A and B exhibit a mixed toxicity relationship.
C. B is safer than A.
D. The therapeutic value of A is greater than B.
E. A is more potent than B.

6. Approximately what percentage of a normally distributed population would be included in ±2 SD?

A. 2%
B. 33%
C. 67%
D. 95%
E. 99%

7. The dose at which three-fourths of the test organisms were observed to exhibit toxicity:

A. LD_{25}
B. LD_{75}
C. LD_{50}
D. TD_{25}
E. TD_{75}

8. List and briefly describe the assumptions related to establishing a dose-response relationship.

9. Define and diagram on the following graph the terms and abbreviations: LD_{10}, LD_{50}, threshold dose, and NOEL.

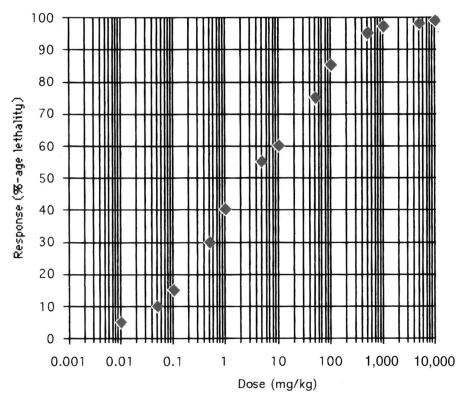

10. Toxicant A has already been plotted on the following graph. Plot data for toxicant B on the graph, then answer the following questions.

A. Based on the LD_{10}, which toxicant is more potent?
B. Based on the LD_{50}, which toxicant is more potent?
C. At what dose do these two toxicants have the same percentage lethality?

Dose (mg/kg)	Toxicant A (%-age lethality)	Toxicant B (%-age lethality)
0.01	10	5
0.05	15	10
0.1	20	15
0.5	30	30
1	35	40
5	45	55
10	50	60
50	60	75
100	65	85
500	85	95
1,000	90	97
5,000	95	98
10,000	97	99

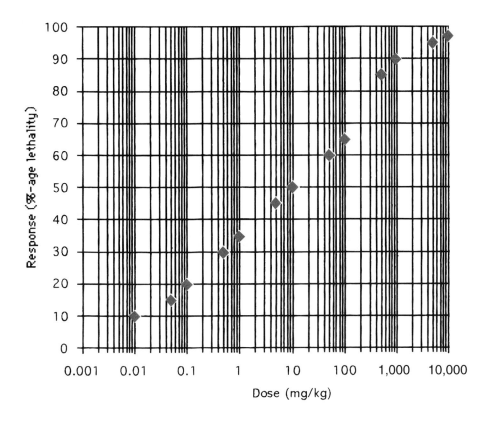

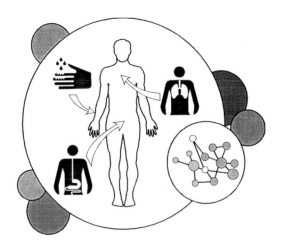

O bjectives

- Describe the ways in which toxicants interact with cells

- Recognize how the molecular characteristics of toxicants affect entrance into a cell

- Explain human anatomy as related to integumentary, respiratory, and digestive systems

- Summarize integumentary, respiratory, and digestive routes of toxicant absorption

K eywords

absorption
active transport
adenosine triphosphate (ATP)
alveolar region
cell membrane
concentration gradient
digestive system
endocytosis
epithelial cells
epithelium
exocytosis
facilitated diffusion
hydrophilic
hydrophobic
integumentary system
lipid soluble
lipophilic
minute volume respiration (MVR)
mucociliary escalator
nasopharyngeal region
nonpolar
occluding cell junctions
Overton's Rules
partition coefficient
percutaneous
phagocytosis

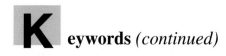

Keywords *(continued)*

phospholipid

phospholipid bilayer

pinocytosis

pneumocytes

Poiseuille's Law

polar molecules

respiratory system

semipermeable membrane

simple diffusion

tracheobronchial region

Interaction of Toxicants with Cells

In general, toxicants exert their effects when they interact with cells. This cellular interaction may occur: (1) on the surface of the cell, (2) within the cell, or (3) in the underlying tissues and extracellular (interstitial) space (Figure 4-1). Although there are four different types of tissues in the human body (epithelial, connective, muscle, and nerve), it is a collection of epithelial cells that forms the epithelium that covers the surface of the body (i.e., skin), and forms the lining of the lumen (or inside walls) of the respiratory and digestive systems.

Chemical characteristics of both the toxicant and cell membrane determine whether any interaction occurs on the surface of the cell or whether the barrier will be effective in keeping the toxicant out of the organism. Under normal conditions, the contacts between adjacent epithelial cells will not permit the passage of substances. This is due to the presence of occluding cell junctions, which are formed by intramembranous proteins arranged to "stitch" the membrane of adjacent cells together.

Cell Membrane Structure

A fundamental question in biology is: How do substances get into and out of a cell? Since the cell is the ultimate level at which toxicants have their impact, the answer to this question is relevant: not only for toxicants, but also for substances needed for cell survival, such as nutrients (glucose), gases (O_2, CO_2), electrolytes (Na^+, K^+, Cl^-), and molecules produced by the cell for export (enzymes, hormones, structural proteins).

A knowledge of the structure of epithelial cells, particularly the characteristics of the cell membrane, or plasma membrane, is important when considering why some toxicants move through the barrier with relative ease while other toxicants find entrance into the body difficult or impossible.

The cell membrane is composed of phospholipid molecules (Figure 4-2). As the term phospholipid implies, there are two components to the molecule: phosphates and lipids. The phosphate head is a region that is hydrophilic. This means that this portion of the molecule prefers associating with water (*hydro-*, water; *-philic*, attraction for or love of). In contrast, the lipid tail is a hydrophobic region, which is repelled by water. Additionally, this region is said to be lipophilic, or attractive to lipid-soluble substances.

The cell membrane is like a sandwich, composed of two layers of phospholipid molecules. A typical cell membrane is about 10 nm (10 one-billionths of a meter) thick. The term phospholipid bilayer provides a good description of the appearance of the "sandwich." In the middle of the phospholipid bilayer is a region where adjacent lipid tails are located. The phosphate heads are found on the inner and outer surfaces of the cell membrane where these regions are exposed to water, the ubiquitous solvent found in living organisms. The phospholipid bilayer forms a semipermeable membrane that surrounds the cell. The membrane is termed semipermeable since it permits some molecules to move across, while at the same time stopping or

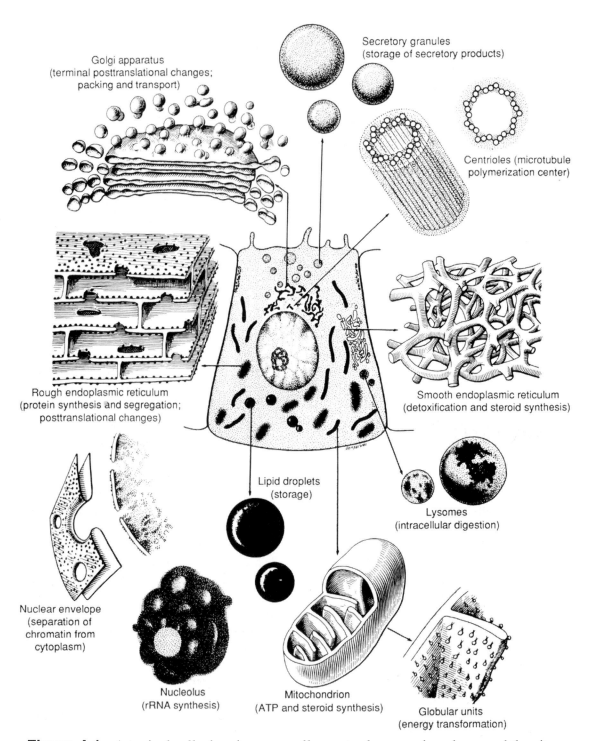

Figure 4-1. A typical cell, showing organelles, cytoplasm, and nucleus, and the sites where toxicants have their impact. (From L. C. Junqueira, J. Carneiro, and R. O. Kelly, *Basic Histology*, 7th edition. Appleton & Lange, 1992. Reprinted by permission.)

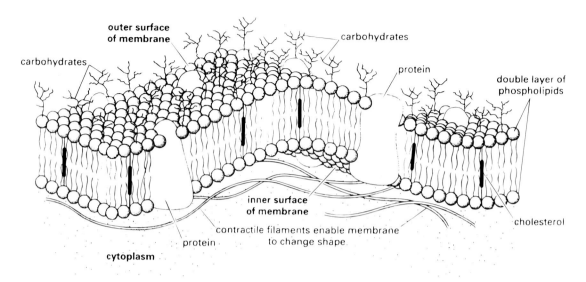

outer surface
of membrane

carbohydrates

carbohydrates

protein

double layer of
phospholipids

inner surface
of membrane

contractile filaments enable membrane
to change shape

protein

cholesterol

cytoplasm

Figure 4-2. Cell membrane structure. (From B. Davey and T. Halliday, editors, *Human Biology and Health: An Evolutionary Approach.* Open University Press, 1994. Reprinted by permission.)

impeding the progress of other molecules.

Processes of Cellular Absorption

Substances use a number of different passive (spontaneous) and active (energy-requiring) transport mechanisms to gain entrance into a cell. The most commonly used process is simple diffusion. In this process the molecule relies on its concentration gradient to enter the cell. This process is passive, as opposed to active, since no cellular energy is used to "power" the toxicant across the cell membrane. Remember, diffusion is the movement of a substance (in this case a toxicant) from a region of high concentration into a region of low concentration and, depending on the direction of the concentration gradient, substances will continue to move into or out of a cell until equilibrium is reached. In the absence of a concentration gradient, no net movement of substances will occur.

A second process used to cross the cell membrane is facilitated diffusion. In this process molecules become bound to specific carrier proteins found on the outer surface of the cell membrane. The molecule is then passed, or passively transported, by the membrane protein into the cell. The "energy" for transport is derived from the potential energy stored in the concentration gradient, not from cell energy input. Facilitated diffusion is a well-known mechanism in the transport of nutrients, such as glucose, across the cell membrane. Facilitated diffusion is thought to play only a minor role in the transport of toxicants into the cell; however, it does serve as an important transport mechanism for the elimination of toxicants or their metabolites from the cell following absorption via other mechanisms.

Third, active transport, as a means to enter the cell, involves the consumption of cellularly-produced energy, such as adenosine triphosphate (ATP). Whereas passive and facilitated diffusion make good use of concentration gradients, active transport enables the cell to transport molecules against, or up, their concentration gradient. Although not a major route of toxicant entry into the cell, active transport, like facilitated diffusion, does play a vital role in the elimination of toxicants or their metabolic intermediates from a cell.

It is impossible for many large molecules and particulates to go into or leave the cell via passive or active transport mechanisms. Instead, these macromolecules enter and exit the cell by two different processes called endocytosis and exocytosis, respectively. During endocytosis the cell membrane will flow around and engulf the macromolecules that are in close proximity to the cell. Once engulfed, the particulate, now with its cell membrane covering, will invaginate or turn inward to form a vesicle. The vesicle will detach from the adjacent cell membrane and become part of the cytoplasm. Phagocytosis (cellular eating) and pinocytosis (cellular drinking) are two types of endocytosis. Phagocytosis, performed by special white blood cells, is responsible for removing particulates from the small sacs (alveoli) in the lung.

Cellular Uptake of Toxicants

Two features characterize toxicants that make use of simple diffusion to enter a cell. First, they are nonpolar or lipid soluble. Nonpolar means that they have a neutral molecular charge distribution, unlike polar molecules, which have a positive or negative charge. And lipid soluble indicates that the toxicant

will dissolve in lipids, such as would be found in the middle of the phospholipid bilayer. Second, they have low molecular weights—that is, they are small, usually with a molecular weight of less than 600 (MW < 600). There is an inverse relationship between the molecular weight of a chemical and its ability to move through the cell membrane. Typically, a nonpolar, lipid-soluble toxicant will diffuse across a cell membrane much more rapidly than a polar, water-soluble toxicant of the same size.

Overton's Rules, as applied to toxicants, summarize the general relationship between polarity and solubility: (1) the permeability of cell membranes to small, nonpolar molecules is directly proportional to the lipid solubility of the toxicant; and (2) the permeability of cell membranes to polar molecules is inversely proportional to the molecular size of the solute. Water, with its small molecular size and high polarity, is an obvious exception to these rules since it readily crosses the cell membrane.

It is often useful when comparing the absorptive behavior of toxicants to determine their relative solubility in lipids and also in water. This is referred to as the partition coefficient, and is defined as the ratio of the toxicant's solubility in a nonpolar solvent, such as chloroform ($CHCl_3$), hexane (C_6H_{14}), or octanol ($C_8H_{17}OH$), to its solubility in the ultimate polar solvent, water (H_2O).

Recognize that toxicants don't always enter a cell to exert their toxic influence. Many toxicants are capable of interacting with other molecules associated with the outer cell membrane surface, such as receptor proteins, recognition proteins, channel proteins, transport proteins, and electron transfer proteins (Figure 4-2). Those toxicants that do enter the cell can potentially interact with a number of different cellular components, including the nucleus, organelles, cytoskeletal proteins, and the cytoplasm, to produce functional and structural anomalies that lead to toxicity (Table 4-1).

Routes of Absorption

Absorption is the process by which toxicants cross the epithelial cell barrier. Depending on the nature of the toxicant, dose, duration, and type of exposure, a toxicant may limit its contact to the outer surface of the epithelial cell barrier, or cross the cell membrane, enter the cell, and possibly move completely through the cell and into the underlying lymphatic or cardiovascular divisions of the circulatory system. There are three primary routes of absorption: (1) percutaneous (integumentary system), or through the skin; (2) the respiratory system; and (3) the digestive system. Under accidental circumstances where there is an unnatural interruption to the integrity of the barrier (e.g., lacerations, punctures, chemical or electrical burns), absorption can also take place in the exposed tissues found beneath the epithelium, such as subcutaneous fat and muscle.

Each route of absorption has its own special type of epithelial cells that unite to form specific tissues. These unique cell and tissue characteristics present the potential toxicant with a different set of structural and functional features that must be overcome to gain entrance into the body. There are also route-specific structures, such as hair follicles in the skin, that actually facilitate the absorption of some toxicants. An awareness of

Table 4-1. Major cellular structures and their functions

Cell structure	Function
Cell membrane	Serves as the cell boundary. May contain protein, glycoprotein, and glycolipid molecules that function as membrane receptors, in membrane transport, and in cell identification. Regulates what enters and leaves the cell.
Nucleus	Directs protein synthesis. Regulates cell metabolism, growth, and reproduction.
Ribosomes	Synthesize proteins for use by the cell or for export.
Mitochondria	Synthesis of ATP; the cell's power plant.
Endoplasmic reticulum (ER)	Two forms are present: smooth ER, which produces lipids and some carbohydrates; and rough ER, with attached ribosomes, involved in protein synthesis.
Lysosomes	Cellular digestion.
Golgi apparatus	Synthesizes carbohydrates, which are packaged with proteins, for cellular use and export.

anatomical and physiological characteristics associated with each route of
absorption is important as a first step in
understanding how toxicants enter the
body.

Percutaneous Route

The integumentary system is the
largest organ system in the human body.
Although the skin is the most obvious
organ of the integumentary system, this
system also includes hair, fingernails
and toenails, and mammary glands.
Skin plays an important role to (1) provide a barrier against the entrance of
toxicants, (2) protect against the harmful effects of ultraviolet radiation, (3)
prevent the entrance of microorganisms, (4) assist in the biotransformation
or metabolic detoxification of toxicants,
(5) eliminate toxicants or their metabolites via sweat or other glandular secretions, (6) regulate body temperature,
and (7) house sensory receptors for temperature, pressure, and pain.

The complexity of skin becomes
apparent when you consider that each
square centimeter (cm^2) of skin contains
approximately 150 nerve endings, 80
sweat glands, 40 sensory receptors, and
15 oil glands, all of which require a supply of blood provided by a meter's
length of small blood vessels.

Skin is composed of the epidermis,
dermis, and hypodermis (Figure 4-3).
The epidermis is an outer protective
region in which the pigment layer
(melanocytes), stratum germinativum,
and stratum corneum are located. The
stratum corneum is typically 15–20
cells thick but will vary according to
race, age, sex, physical state of individual, climatic changes, and other factors.
New epithelial cells arise in the stratum

germinativum and migrate outward to
become the stratum corneum. Together
these two cell zones form stratified
squamous epithelium. The hardened or
keratinized stratum corneum, along
with extracellular lipids, provides the
main protective barrier to water loss or
entrance and to toxicant entrance.

The dermis is composed of connective tissue that is highly vascular (having numerous blood vessels). It is in the
dermis that oil and sweat glands, hair
follicles, and sensory receptors are
found. The underlying hypodermis is
composed of connective tissue and adipose tissue, where about half of the
body's fat storage is found. This fatty
region is often the site of lipid-soluble
toxicant storage.

The combined thickness of the epidermal and dermal zones varies in different areas of the body. In areas that
receive constant abrasion, such as the
elbows, knees, and palms, skin may be
1 mm thick, whereas in less exposed
areas like the eyelid and antecubital
space (in front of the elbow), skin is
only 0.2 mm thick.

Several routes of absorption are
possible through the skin. The most
common is the cutaneous adsorption of
a toxicant followed by passive diffusion
through the epidermis into the dermis
where the toxicant might enter a blood
vessel. Passage into the dermis is
enhanced if the toxicant enters a sweat
gland or hair follicle. Since these structures originate in the dermis and penetrate through the epidermis, this route
effectively bypasses the protective barrier provided by the epidermis.

How quickly a toxicant diffuses
through the epidermis is affected by a
number of factors, including dose,
length of exposure, lipid solubility, and

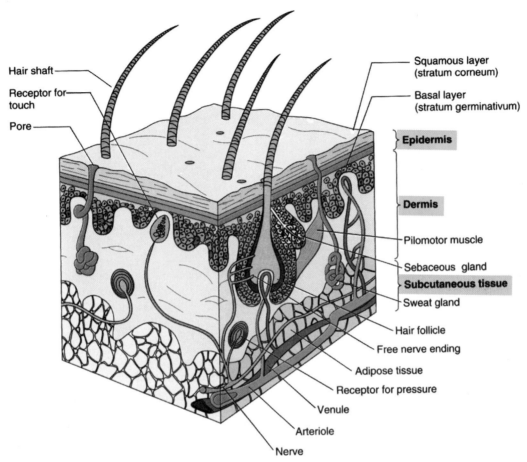

Figure 4-3. Diagram of human skin. (From M. C. Willis, *Medical Terminology: The Language of Health Care*. Williams & Wilkins, 1996. Reprinted by permission.)

skin location. Toxicants that are small, nonpolar, and lipid-soluble will diffuse most rapidly. Diffusion is also accelerated when the skin has been pretreated with organic solvents, such as chloroform ($CHCl_3$), methanol (CH_3OH), or dimethyl sulfoxide ($[CH_3]_2SO$). Enhanced epidermal permeability is thought to result from the removal of extracellular lipids by these solvents.

Finally, when toxicants become localized in the epidermis, local toxicity, rather than systemic toxicity, is the likely result. This is because the epidermis is avascular (having no blood ves-

sels). Without a transport mechanism, toxicants cannot be distributed to other areas of the body where systemic toxicity may result.

Respiratory System Route

The respiratory system is composed of the nasopharyngeal, tracheobronchial, and pulmonary anatomical regions (Figure 4-4). Each region contributes a unique functional component that prohibits or limits the ability of toxicants to enter the body. In addition to the tissue type previously noted in the epidermis of

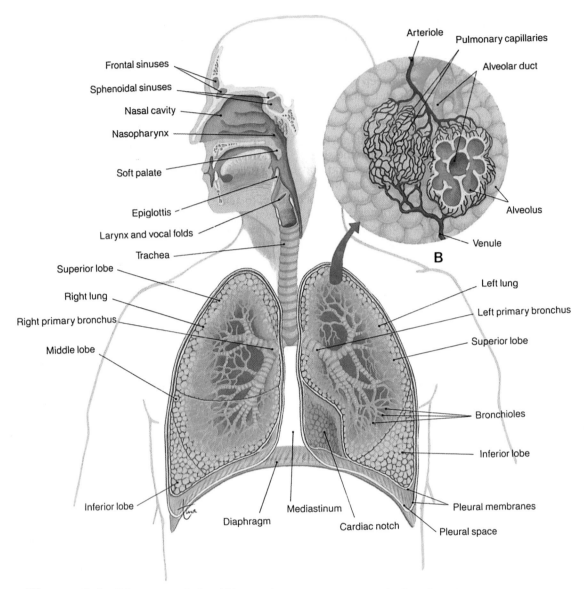

Figure 4-4. Diagram of the (A) respiratory system and (B) alveolus. (From V. C. Scanlon and T. Sanders, *Essentials of Anatomy and Physiology*, 2nd edition. F. A. Davis, 1995. Reprinted by permission.)

the skin (i.e., stratified squamous epithelial tissue), the lumen of the respiratory system also contains ciliated columnar epithelium and ciliated cuboidal epithelium, the appearance of which resembles the names. Unlike in skin, the stratified squamous epithelium in the respiratory system is nonkeratinized, and it becomes less frequent farther down the respiratory passageways. Due to its nonkeratinized form, the stratified squamous epithelium in the respiratory system is less effective in serving as a barrier. Mucus-secreting cells, smooth muscle, cartilage, and immune cells can be found in specific regions of the respiratory system.

Anatomically, the nasopharyngeal region includes the nares (nostrils), nasopharynx, oropharynx, laryngopharynx, and larynx. It is in this region that air first enters the respiratory system. As air moves through the nasopharynx it is cleaned, humidified, and thermally adjusted. Hairs and mucus in this region are effective in trapping particulates greater than 5 μm, in diameter, which prevents them from entering the lower regions of the respiratory system.

The tracheobronchial region is composed of the trachea (or windpipe), bronchi (singular bronchus), and bronchioles. The two bronchi result from the initial bifurcation (split) of the trachea. The bronchioles represent additional bifurcations, with the smallest bronchioles being the result of about 16 separate bifurcations. These highly branched (bifurcated) and narrow air passageways increase the available surface area upon which toxicants can interact. Mucus on the luminal surfaces of these cells is effective in trapping small particulates (2–5 μm in diameter) and water-soluble toxicant gases.

In the tracheobronchial region the importance of the cilia on the columnar epithelial cells becomes evident. Each cell may contain over 200 cilia and each individual cilium is only 6–10 μm long. With coordinated motion these cilia move to "sweep" the mucus, along with trapped particulates and gases, up and away from the delicate alveolar tissues where gas exchange takes place. This combined action of the mucus and cilia is called the mucociliary escalator. If a toxicant impairs the normal functioning of the mucociliary escalator, substances trapped in the mucus will no longer be transported out of the lower regions of the respiratory system. The consequences will be prolonged exposure of the epithelium to toxicants, accumulation of mucus in the respiratory passageways (which decreases the cross-sectional area), and possibly impaired function of the alveoli. Keep in mind that small changes in the cross-sectional area of the respiratory passageways leading to the alveolar region profoundly reduce the flow of air (i.e., to the 4th power), as defined by Poiseuille's Law:

$$V \, dt = flow = \frac{(\text{pressure gradient})(\text{radius of passageway})^4}{(\text{length}) \, (\text{viscosity})}$$

In the alveolar region, small terminal bronchioles give rise to respiratory bronchioles and their associated alveoli (singular alveolus), which continue to bifurcate an additional six or seven times. The end result is 400–1,200 million alveoli in healthy adult human lungs. Although the additional bifurcations do not decrease the cross-sectional area of the passageways, they do significantly increase the surface area across which gas exchange takes place to about 70–80 m^2. It is possible for non-water-soluble gases to reach the alveoli, as will particulates less than 1 μm in diameter. Once in the alveoli the toxicant may initiate an immune response involving phagocytizing white blood cells, interact with the surface of, or enter, special lung cells called pneumocytes, or pass through pneumocytes to enter the cardiovascular system or interstitial (extracellular) space.

In addition to factors such as chemical characteristics of the toxicant, dose, and length of exposure, the amount of a toxicant that can enter the body using the respiratory route will depend on a parameter unique to the respiratory system—the minute volume respiration (MVR). The MVR is the tidal volume (amount of air breathed in on each respiratory cycle) multiplied by the respirations per minute. For example, under normal conditions about 0.5 L is inhaled with each respiration and under resting conditions a normal adult will breathe 12 times per minute, in which case the MVR = 0.5 L/resp × 12 resp/min = 6.0 L/min. Under conditions of physical exercise the MVR can rise significantly; this in turn will increase contact of the toxicant with the respiratory tissues.

The respiratory system, with its close anatomical and physiological association with the cardiovascular system, is one of the prime sites for the absorption and distribution of toxicants. The cells lining the lumina of the respiratory system are highly susceptible to toxicants, both particulates and gases. Pneumocytes, which form the delicate alveoli, are capable of rapidly transporting toxicants directly into the pulmonary blood circulation for distribution to the rest of the body.

Digestive System Route

The digestive system includes the mouth, oral cavity, esophagus, stomach, small intestine, large intestine, rectum, and anus, as well as accessory organs such as the pancreas and liver (Figure 4-5A). Four distinct zones are found in the digestive system: (1) mucosa, (2) submucosa, (3) muscularis, and (4) serosa (Figure 4-5B and C). Depending on the location in the digestive system, the mucosa lining the lumen can be a tough, abrasion-resistant, nonkeratinized, stratified squamous epithelium (as in the mouth and esophagus) or a simple columnar epithelium (as in the small intestine), which functions well in absorption and secretion. The mucosa is avascular and in some regions (small intestine) has numerous projections called villi, each of which will have about 2,500 microvilli protruding from its surface (Figure 4-5D). Again, as previously evidenced in the highly branched alveolar region of the lung, the villi and microvilli serve to increase the absorptive surface area of the small intestine.

The submucosa contains abundant blood vessels, lymphatics, and nerves. In this region toxicants that have entered through the mucosal barrier can enter the blood supply to be transported to other body regions. The muscularis is the third layer inward from the luminal surface. This layer contains the involuntary (or smooth) muscle that produces the rhythmic peristaltic contractions that mix and move food through the digestive system. The outermost covering of the digestive tube is called the serosa. Composed of fibrous connective (or collagenous) tissue, the serosa serves to encase the other three zones.

Absorption can take place across the mucosal lining anywhere along the entire length of the digestive system. However, the time food and potential toxicants are in the mouth and esophagus is usually too short to be a major site of toxicant entry. The stomach, where food may remain for about 2 hours, is the site where mechanical digestion occurs and where hydrochloric acid (HCl), gastric enzymes, and bacteria help to chemically break down food.

Most absorption of food and toxicants takes place in the small intestine. Pinocytosis is one way substances can enter the systemic circulation via the lymph vessels, which are abundant in the submucosa that underlies the mucosa. This route of absorption is not as direct as it may seem. Toxicants must enter and move through the mucosal epithelium

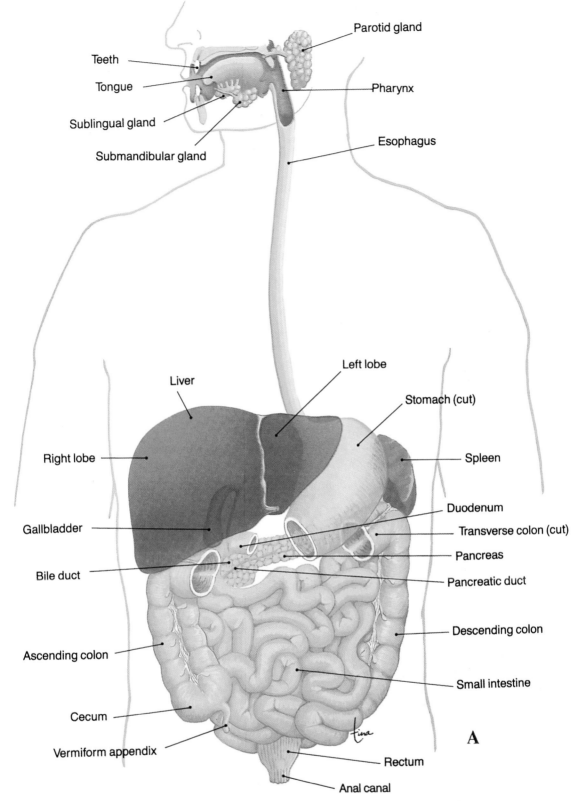

Figure 4-5. Diagram of the digestive system (A), and general anatomic structures as viewed in cross section of (B). (From V. C. Scanlon and T. Sanders, *Essentials of Anatomy and Physiology*, 2nd edition. F. A. Davis, 1995. Reprinted by permission.)

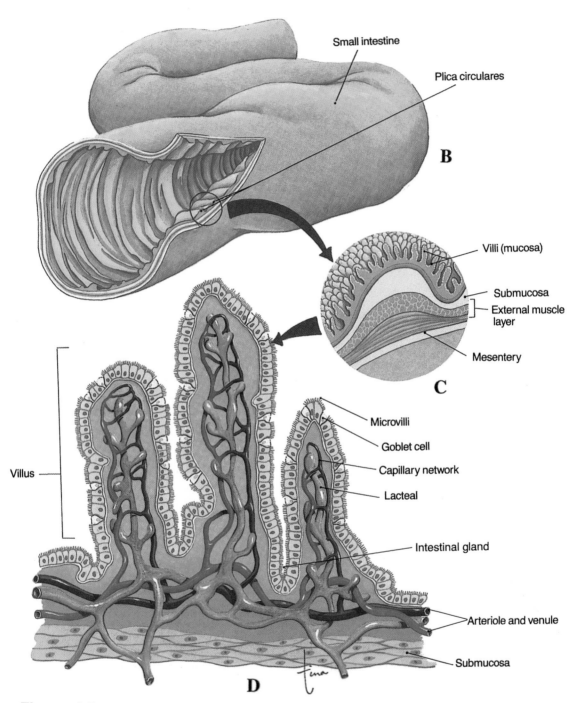

Figure 4-5. (*Continued*) (B) The small intestine, (C) intestinal wall, and (D) villi. (From V. C. Scanlon and T. Sanders, *Essentials of Anatomy and Physiology*, 2nd edition. F. A. Davis, 1995. Reprinted by permission.)

and continue into the submucosa, where they enter the lymphatic system. Lymph is then transported through lymph ducts and vessels up to the region of the heart, where two large lymph vessels, the common thoracic duct and right lymphatic duct, drain into the cardiovascular system near the aorta. Similar lymphatic transport may also occur with the integumentary and respiratory routes of absorption.

The large intestine is the final region of the digestive system. This region lacks villi and is not considered a major site for absorption of toxicants. However, this region does function to remove liquid from chyme (food that is mixed with digestive "juices" such as HCl and enzymes), thereby producing the more solid form of indigestible waste called feces.

Review Questions

1. Where do toxicants exert their effects in the body?

A. On the surface of cell membranes.
B. Within cells.
C. In spaces between cells.
D. A and B
E. A, B, and C

2. All of the following are true statements about cell membranes except:

A. They are composed of phospholipid molecules.
B. The phosphate "head" is the hydrophilic region.
C. The lipid tails are found on the inner and outer surfaces of the phospholipid bilayer.
D. The middle of the phospholipid bilayer is lipophilic.
E. The phospholipid bilayer forms a semipermeable membrane.

3. Which transport mechanism relies only on a concentration gradient to enter the cell?

A. Active transport
B. Endocytosis
C. Phagocytosis
D. Pinocytosis
E. Simple diffusion

4. Defined as the ratio of the toxicant's solubility in a nonpolar solvent to its solubility in water.

A. Absorption
B. Concentration gradient
C. Facilitated diffusion

D.   Overton's Rules
E.   Partition coefficient

5.   Diffusion of a toxicant through the epidermis is affected by:

A.   Length of exposure
B.   Lipid solubility of toxicant
C.   Skin location
D.   A and B
E.   A, B, and C

6.   Which is *not* a true statement about the tracheobronchial region of the respiratory system?

A.   It is composed of trachea, bronchi, bronchioles, and alveoli.
B.   Columnar epithelial cells lining this region contain cilia.
C.   Mucus in this region is effective in trapping small particulates (2–5 μm in diameter).
D.   It is characterized by the mucociliary escalator.
E.   The smallest of bronchioles in this region are the result of about 16 separate bifurcations.

7.   Which route of entry involves crossing the mucosal zone prior to entering the submucosal zone, which contains abundant blood vessels, lymphatics, and nerves?

A.   Digestive system route
B.   Percutaneous route
C.   Respiratory system route
D.   A and B
E.   A and C

8.   The absorption of toxicants by the respiratory system is affected by:

A.   Chemical characteristics of the toxicant
B.   Length of exposure
C.   Minute volume respiration
D.   A and B
E.   A, B, and C

9.   Diagram the anatomical structures associated with the route of absorption in each of the following: percutaneous, respiratory system, and digestive system.

10.   What molecular characteristics would the ideal toxicant possess to maximize entrance into the body?

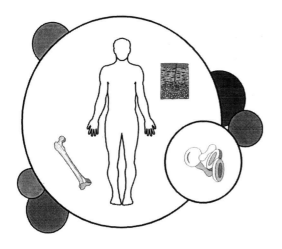

bjectives

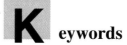

eywords

- Identify the ways toxicants are distributed in the body

- Recognize the relationship between a specific route of absorption and the related pathways for distribution of a toxicant

- Describe the factors affecting distribution of toxicants to tissues

- Define volume of distribution

- List the sites for toxicant storage

- Discuss how storage influences the half-life of a toxicant

adipose tissue
albumin
arterial vessels
blood flow
blood flow/mass ratio
blood plasma
blood-brain barrier
capillaries
cardiac output
distribution
erythrocytes
heart
interstitial fluid
intracellular fluid
leukocytes
lymph
lymph capillaries
lymph nodes
lymphatic system
lymphocytes
lymphoid tissue
placental barrier
plasma protein
platelets
portal vein
storage
venous vessels
volume of distribution (V_D)

Distribution of Toxicants

Following exposure, the fate of a toxicant in the body will be affected by many processes. Recall that toxicokinetics is the study of five processes: (1) absorption, (2) distribution, (3) storage, (4) biotransformation, and (5) elimination.

Distribution occurs when a toxicant is absorbed and subsequently enters the lymph (L. *lympha*, clear water) or blood supply for transport to other regions of the body. The lymphatic system is a part of the circulatory system and drains excess fluid from the tissues (Figure 5-1). Included in the lymphatic system are lymph capillaries, lymph nodes, aggregations of lymphoid tissue (tonsils, spleen, and thymus), and circulating lymphocytes (one of the five different types of white blood cells).

The cardiovascular part of the circulatory system includes the heart, arterial vessels, venous vessels, capillaries, and the circulating medium called blood (Figure 5-2). Blood is composed of three cellular components: (1) erythrocytes (red blood cells or RBCs), (2) leukocytes (white blood cells or WBCs), and (3) platelets (thrombocytes), all of which are suspended in a yellowish, noncellular fluid called blood plasma. Although both the lymphatic and cardiovascular circulatory systems are capable of distributing toxicants, the blood in the cardiovascular system is responsible for most transport.

A number of features must be considered when determining whether or not a toxicant will be distributed to tissues in distant regions of the body. These features include duration of exposure, dose, the chemical characteristics of the toxicant, and the presence of lymphatic or blood vascular components. The precise location where the toxicant enters the bloodstream is important because once a toxicant gains entrance into the organism, the other toxicokinetic processes—such as storage, biotransformation, and elimination—will affect the concentration of the toxicant in the lymph or blood. When these toxicokinetic processes occur soon after the entrance of the toxicant into the blood supply, or immediately "downstream" from the point of entry, then blood levels of the toxicant may be diminished or eliminated, thus reducing or eliminating toxicity.

For example, toxicants absorbed through the digestive system may enter the cardiovascular circulatory system directly or they may take a circuitous route involving lymphatic circulation. Toxicants that directly enter the bloodstream in the highly vascularized submucosa of the digestive system are initially transported by the portal vein to the liver, a primary site for biotransformation or metabolic detoxication. Blood from the liver is transported to the heart, where it first enters the pulmonary circulation and then is pumped to the systemic circulation.

Toxicants that enter the lymphatics associated with the digestive system move with the lymph through afferent (prenodal) lymph vessels to lymph nodes, then on to efferent (postnodal) lymph vessels and larger vessels called lymph trunks. The two lymph trunks, right and left, ultimately empty into the venous blood supply associated with the right and left jugular veins, respectively, then return to the heart to be pumped through the systemic circulation.

Toxicants absorbed through the respiratory system may enter directly into

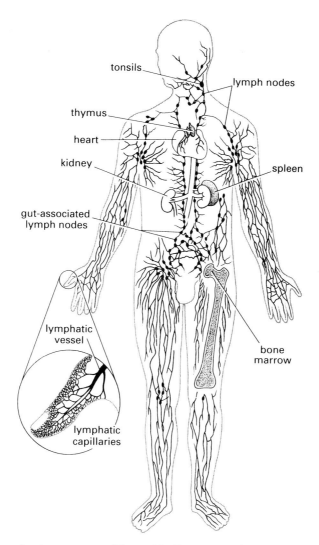

tonsils

lymph nodes

thymus

heart

kidney

spleen

gut-associated
lymph nodes

lymphatic
vessel

bone
marrow

lymphatic
capillaries

Figure 5-1. Lymphatic system. (From B. Davey and T. Halliday, editors, *Human Biology and Health: An Evolutionary Approach.* Open University Press, 1994. Reprinted by permission.)

the pulmonary circulation and, along with oxygen, may be transported from the lungs to the heart via the pulmonary vein. The heart pumps the toxicant and oxygenated blood into the systemic circulatory pathway, where it is distributed to the tissues of the body. Other inhaled toxicants, such as particulates, move into the spaces between cells, where they are transported at a slow pace along with the lymph in lymphatic vessels.

Finally, toxicants absorbed through the skin may enter the peripheral blood supply for distribution to tissues far removed from their original route of entry into the body. The toxicant may then interact with these distant cells to produce toxicity. Keep in mind that additional processes that affect toxicokinetics, such as storage, biotransformation, and elimination, may lower the concentration of the toxicant.

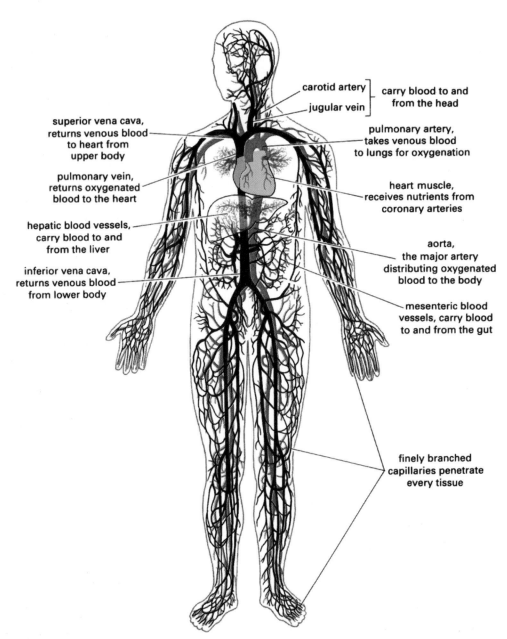

Figure 5-2. Circulatory system. (From B. Davey and T. Halliday, editors, *Human Biology and Health: An Evolutionary Approach.* Open University Press, 1994. Reprinted by permission.)

Factors Affecting Distribution of Toxicants to Tissues

The distribution of toxicants to tissues is dependent on several factors: (1) physical and chemical properties of the toxicant, (2) concentration of the toxicant in the blood and in the tissues (i.e., concentration gradient), (3) volume of blood flowing through a specific tissue, (4) tissue specificity or preference of the toxicant, and (5) presence of special "barriers" to slow down toxicant entrance.

Properties of the Toxicant

The same physical and chemical characteristics that determined a toxicant's initial absorptive behavior also determine its likelihood of being distributed to tissues in the organism. Distribution to tissues requires the toxicant to once more interact at the level of the cell. This time, instead of interacting with the cells that form the "absorptive barrier," its interaction is with the cells that form the tissues of the target organ. Molecular weight and polarity are again of concern, with the smaller, nonpolar toxicants gaining preferential entrance.

Concentration Gradient

The concentration gradient between the amount of the toxicant in the blood as compared to the tissues will depend on a number of features. Following absorption a toxicant will be diluted by the fluid volume present in the organism. This fluid is found in three separate sites: (1) blood plasma, which accounts for about half of the blood volume—depending on the individual's sex and age, blood volumes in humans range from 4 to 6 L, and typically will account for 7–9% of the total body weight; (2) interstitial fluid volume, the fluid between cells, which makes up about 13% of the total body weight; and (3) intracellular fluid, the fluid inside cells, which accounts for about 40% of the total body weight. Dilution, and the resulting concentration of a toxicant in any of these three sites, is an important consideration, since the passive diffusion of the toxicant into or out of these sites will be determined by the toxicant's concentration gradient. The apparent volume of distribution (V_D) is used to define the volume of body fluids in which a toxicant is distributed. The following expression is often used:

$$V_D (L) = \frac{dose\ (mg)}{plasma\ concentration\ (mg/L)}$$

The apparent V_D does not indicate in which of the three fluid volume sites a toxicant is distributed. Hence, if a toxicant is distributed only in the plasma fluid, a low V_D would result; however, if a toxicant is distributed in all sites (blood plasma, interstitial and intracellular fluids) the V_D will be greater. The validity of the apparent V_D can be further compromised by toxicants that undergo rapid storage, biotransformation, or elimination. Additionally, some toxicants bind to plasma proteins (e.g., albumin) circulating in the blood. Binding to plasma proteins "removes" the toxicant from added interaction with cells. This removal effectively reduces the concentration of the toxicant in the plasma or V_D, since only "free," non-protein-bound toxicants will interact with cells.

Blood Flow

The volume of blood flowing through specific tissues or organs in the human body is an important factor affecting distribution. Two parameters affect the accumulation of a toxicant in an organ: (1) the volume of blood flowing through the organ, and (2) the size or mass of the organ. A combination of these two, the blood flow/mass ratio, permits the comparison of toxicant accumulations in different organs (Table 5-1). The percentage of the total cardiac output received by each of the body organs is also important, since organs that receive larger blood volumes can potentially accumulate more of a given toxicant. Cardiac output equals the volume of blood pumped per heart beat (stroke volume) times the heart rate (beats per minute). During exercise, cardiac output increases as a result of a combination of a larger stroke volume and an increase in heart rate.

Body regions that receive a large percentage of the total cardiac output or have high blood flow/mass ratios (i.e., mL/kg) include the liver, kidneys, heart muscle, and brain. Blood flow to the brain is constant, even during exercise. Approximately half of the total cardiac output will be sent to the liver (27.8%) and kidneys (23.3%). This is not surprising, as these are the major organs involved in the elimination of toxicants or their metabolites. Skeletal muscle and skin have intermediate blood flow/mass ratios. However, during strenuous exercise blood flow to skeletal muscles may increase from less than 750 mL/min to more than 10,000 mL/min (or 10 L/min). This increase in blood flow may subject skeletal muscles and other organs to greater toxicant exposure and accumulation.

Bone and adipose tissues have relatively low blood flow/mass ratios. This is important to remember, since these areas also serve as primary storage sites for many toxicants, especially those that are fat soluble as well as those that readily associate (or complex) with minerals commonly found in bone.

Affinity of Toxicants for Specific Tissues

Some tissues are "attractive" to specific toxicants and, in spite of the rather

Table 5-1. Comparison of mass, blood flow, and percentage of total cardiac output for selected body regions

Region	Mass (kg)	Blood flow (mL/min)	Blood flow (mL/100g/min)	Total cardiac output (% of total)
Liver	2.6	1500	58.0	27.8
Kidneys	0.3	1260	420.0	23.3
Skeletal muscle	31.0	840	2.7	15.6
Brain	1.4	750	54.0	13.9
Skin	3.6	462	12.8	8.6
Heart muscle	0.3	250	84.0	4.7
Other body	23.8	336	1.4	6.2
Total body	63.0	5400	8.6	100.0[a]

Note. After P. Bard, editor, *Medical Physiology*, 11th edition. C. V. Mosby, 1961.
[a]Column total does not equal 100 due to rounding of region values.

low flow of blood to these tissues, they will preferentially accumulate a given toxicant to disproportionately high concentrations. A good example is adipose tissue, such as subcutaneous fat, which is poorly perfused by blood but is an attractive tissue for lipid- or fat-soluble toxicants. Once deposited in these storage tissues, toxicants may remain for long periods of time, due to their solubility in the tissue and the relatively low blood flow.

Structural Barriers to Toxicant Entrance

During distribution, the passage of toxicants from capillaries into tissues or organs in the body is not uniform. Some organs (e.g., brain, placenta, testes) have specialized barriers that make it difficult for toxicants to diffuse into their cells. For example, the brain is protected by the blood-brain barrier. This barrier is formed by specialized glial cells called astrocytes ("star cells"), which possess many small thread-like branches. Numerous "endfeet" on the branches of each astrocyte attach to the outer surface of the endothelial cells that line the capillaries, forming a barrier that separates the endothelium of the capillary from the neurons of the brain. Lipids in the plasma membrane of astrocytic endfeet impede the diffusion of water-soluble toxicants into the brain. In addition, tight junctions between adjacent endothelial cells limit the passage of water-soluble molecules. Although the term "blood-brain barrier" indicates a rather impenetrable barrier, the "barrier" only serves to slow down the rates at which toxicants cross into brain tissue.

The placental barrier is another impediment to chemical substances. Besides providing for the nutritional, gas exchange, and excretory needs of the developing fetus, the placenta also protects the fetus from toxicants absorbed and subsequently distributed in the maternal circulation. The barrier is formed from cell layers between the maternal and fetal circulatory vessels in the placenta. Lipids present in the plasma membranes of these additional cells serve to limit the diffusion of water-soluble toxicants. As with the blood-brain barrier, the placental barrier only slows down the diffusion of toxicants from maternal circulation into fetal circulation, reducing exposure of developing fetal tissues to toxicants.

Storage of Toxicants

Once distribution occurs, toxicants can undergo other toxicokinetic processes such as storage, biotransformation, and elimination. Storage results when toxicants accumulate in specific tissues or become bound to circulating plasma proteins (Figure 5-3). Both mechanisms reduce the concentration of the "free" toxicant in the blood plasma.

Plasma Protein Storage

Albumin is the most abundant circulating plasma protein and the most common plasma protein to which toxicants are bound. The amount of a specific toxicant that will become bound to plasma proteins varies widely (0–99%). Toxicants bound to plasma proteins are "stored"; even though the toxicant-plasma protein complex may continue to circulate through organs,

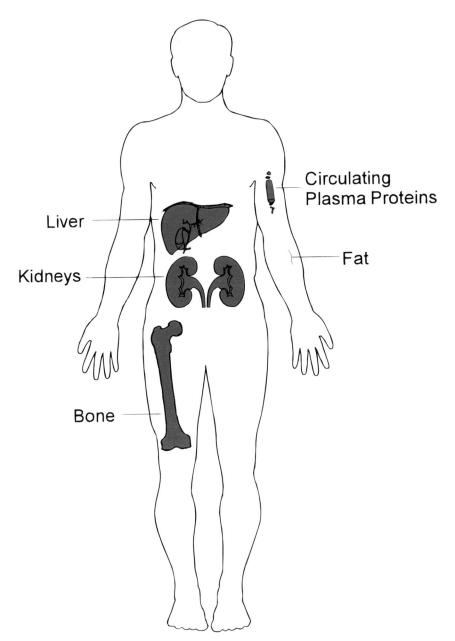

Liver

Kidneys

Bone

Circulating
Plasma Proteins

Fat

Figure 5-3. Common storage locations for toxicants in the body.

the complex will not interact as only the free toxicant can interact with the cell.

Chemical bonding of toxicants to plasma proteins may be accomplished through covalent (i.e., electron sharing) or noncovalent (e.g., ionic) mecha- nisms. Replacement reactions, in which the toxicant is "kicked off" the plasma protein by another molecule with a higher affinity or attraction, are a con- cern. Once "kicked off," the now "free" toxicant will increase its concentration circulating in the blood. If this occurs rapidly it can pose serious problems related to toxicity.

Storage in Bone

Bone is composed of proteins (e.g., collagen, glycosaminoglycans, and proteoglycans) and the mineral salt hydroxyapatite $(Ca_{10}(PO_4)_6(OH)_2)$. Although bone appears to be rather static, it is not—bone is alive, and will bleed since it contains blood vessels. Osteoblasts are the cells responsible for bone formation, and osteoclasts are the cells that reabsorb bone. During bone formation, osteoblasts produce extracellular proteins and also create a chemical environment that favors the natural or biogenic precipitation of hydroxyapatite.

During the normal extracellular processes that lead to the formation of bone, a number of elements and compounds may become involved in chemical substitution reactions. Fluoride (F^-) may be substituted for hydroxyl (OH^-), and strontium (Sr) or lead (Pb) may be substituted for calcium (Ca). Once substituted, elements or compounds become incorporated into the bone matrix. Under normal conditions bone is continually being recycled through resorptive and depositional processes. On an average, the minerals in bone are recycled every 7–10 years—this means that any chemicals locked in the matrix will eventually be released from storage to reenter the circulatory system.

Storage in the Liver

The liver is good at concentrating toxicants—it has a large blood flow/mass ratio, it receives the largest percentage of the total cardiac output, and its hepatocytes (i.e., liver cells) contain cytoplasmic proteins that bind to numerous chemicals, including toxicants. Not only is the liver a primary storage site, but it is also the site where most toxicant biotransformation takes place.

Storage in the Kidneys

The bilaterally placed kidneys have the highest blood flow/mass ratio of all the organs, more than four times greater than that of heart muscle. The large volume of blood flowing though the kidneys preferentially exposes these organs to toxicants. Storage in the kidneys, unlike bone, is not confined to mineral matrices, which have slow turnover rates. Instead, soft tissue associated with the nephron (the functional unit of urine formation) is exposed to the toxicant or its metabolites.

Storage in Fat

The storage of triglycerides (neutral fat) is a major function of adipose tissue. Subcutaneous adipose tissue is the site for about half of all stored neutral fat in the body. Additional fat can be found around the kidneys, in the intestinal omenta (membranes that hold the stomach and intestines in their proper anatomical positions), in the genital areas, between muscles, behind the eyes, and in folds on the surfaces of the heart and intestines. As with storage in bone, the triglycerides deposited in these sites are continually exchanged with the blood and may be redeposited in other adipose tissue cells.

Given that many toxicants are lipophilic, they will readily penetrate cell membranes and become concentrated in adipose tissue. Storage results when toxicants undergo physical dissolution in the neutral fats found in adipose tissue. Once deposited, toxicants may be released, as a result of normal processes of exchange, for distribution to other sites where they may be redeposited, biotransformed, or eliminated.

Review Questions

1. The distribution of toxicants to distant regions of the body is influenced by:

A. Duration of exposure
B. Chemical characteristics of the toxicant
C. Location of toxicant entrance into the bloodstream
D. A and B
E. A, B, and C

2. Which sequence best represents the pathway by which toxicants move through the lymphatic system to enter the bloodstream?

A. Enter lymph → lymph nodes → afferent lymph vessels → efferent lymph vessels lymph trunks → enter venous blood supply
B. Enter lymph → afferent lymph vessels → lymph nodes → efferent lymph vessels lymph trunks → enter venous blood supply
C. Enter lymph → lymph trunks → lymph nodes → afferent lymph vessels → efferent lymph vessels → enter venous blood supply

3. The portal vein is associated with which organ?

A. Lung
B. Liver
C. Kidney
D. Small intestine
E. Spleen

4. The validity of the apparent V_D can be compromised by:

A. Toxicants that undergo rapid storage
B. The rapid elimination of a toxicant
C. Binding of the toxicant to plasma proteins
D. A and B
E. A, B, and C

5. Under normal conditions approximately half of the total cardiac output will be sent to:

A. Bone and adipose tissues
B. Brain and kidneys

C. Kidneys and liver
D. Liver and spleen
E. Skeletal muscle and brain

6. Which is *not* a true statement about the blood-brain barrier?

A. It is formed by special glial cells called astrocytes.
B. The cell membranes of astrocytic "end-feet" impede the diffusion of lipid-soluble toxicants into the brain.
C. Tight junctions between adjacent endothelial cells limit the passage of hydrophilic molecules into the brain.
D. The barrier functions to slow down the rate of toxicant entrance into brain tissue.
E. The barrier separates the circulatory system from neurons in the brain.

7. Storage of toxicants in the body results when toxicants:

A. Accumulate in specific tissues
B. Become bound to circulating plasma proteins
C. Are "kicked off" the plasma proteins by replacement reactions
D. A and B
E. A, B, and C

8. The liver is a primary site of toxicant storage because:

A. It has a large blood flow/mass ratio.
B. Hepatocytes contain cytoplasmic proteins that bind to toxicants.
C. Toxicants are readily incorporated into the bone matrix.
D. A and B
E. A, B, and C

9. Discuss factors that affect the distribution of toxicants to tissues.

10. Diagram the sites of toxicant storage in the body.

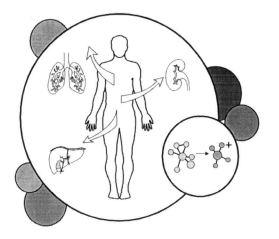

Biotransformation and Elimination of Toxicants

Objectives

- Explain the role of biotransformation in toxicokinetics

- Describe how biotransformation facilitates the elimination of toxicants or their metabolites from the body

- Distinguish between phase I and phase II biotransformation reactions

- Define bioactivation or toxication

- Identify the tissues responsible for biotransformation reactions

- List the factors affecting biotransformation in humans

- Summarize the role of elimination in toxicokinetics

- Describe processes occurring in the kidney, liver, and lung as related to the elimination of toxicants

Keywords

anabolism
bioactivation
biological half-life ($T_{1/2}$)
biotransformation
catabolism
conjugate
conjugation reactions
cytochrome P-450
cytosolic enzymes
detoxication
enterohepatic loop
fecal excretion
glomerular filtration
glucuronidation
hydrolysis
intestinal excretion
ionized molecules
metabolism
microsomal enzymes
nephrons
oxidation
phase I biotransformation
phase II biotransformation
reabsorption
reduction
secretion
sulfate conjugation
toxication

Biotransformation of Toxicants

Recall that lipophilic toxicants, particularly those that are nonpolar and have low molecular weights, are readily absorbed through the cell membrane. These lipophilic toxicants, easily absorbed and potentially systemically distributed, are difficult to eliminate from the body in their original chemical form. The very chemical nature that facilitated their absorption inhibits their elimination—they are lipophilic. With continued absorption and no elimination, many lipophilic toxicants would accumulate within the body. To be eliminated, these lipophilic toxicants must be metabolized or biochemically changed to metabolites (i.e., altered toxicants) that are hydrophilic. In contrast, water-soluble toxicants are generally eliminated from the body in their original chemical form.

Metabolism is the sum of biochemical changes occurring to a molecule within the body. These chemical changes occur within the cell. Anabolism is the sum of biochemical changes that "build up" complex molecules (e.g., proteins). Catabolism is the sum of biochemical changes that "break down" complex molecules (e.g., degradation of glucose). Metabolic processes may take place "free" in the cytoplasm or be "restricted" to specific organelles found within the cell.

The human body has the capacity to eliminate most toxicants either in their original chemical form or as a metabolite. The major routes of elimination (feces and urine) are well known. Biochemical processes or metabolic pathways used by the body to facilitate elimination of toxicants, as well as mol-ecules or their metabolites which occur naturally in the body, are not always appreciated. Biotransformation is the process by which both endogenous (formed within) and exogenous (formed outside) substances that enter the body are changed from hydrophobic to hydrophilic molecules to facilitate elimination from the body.

Biotransformation is responsible for changing naturally occurring lipophilic molecules into hydrophilic metabolites that are more readily eliminated from the body. A good example is the fate of hemoglobin, the oxygen-carrying iron–protein complex in red blood cells. Under normal conditions hemoglobin is metabolized to bilirubin, one of a number of hemoglobin metabolites. Bilirubin is toxic to the brain of newborns and, if present in high concentrations, may cause irreversible brain injury. Biotransformation of the lipophilic bilirubin molecule in the liver results in the production of a water-soluble (hydrophilic) metabolite excreted into bile and eliminated via feces.

Typically biotransformation produces four changes that facilitate elimination of toxicants: (1) the resulting metabolites, or altered toxicant molecules, are chemically distinct from the original toxicant; (2) the metabolites are usually more hydrophilic than the original toxicant; (3) the hydrophilic nature of the biotransformed metabolites reduces their ability to cross membranes, thereby altering its distribution to tissues; (4) there is reduced *re*absorption of metabolites by cells associated with the organs of elimination (kidneys and intestines).

The rate at which a toxicant is removed from the body is dependent on its rate of biotransformation in the body and rate of elimination from the body. The biological half-life ($T_{1/2}$) is the time required to reduce by half the quantity of a toxicant present in the body (e.g., plasma). The biological half-life provides a means for comparing the residence times for different toxicants in the body. This information is useful when establishing "safe" exposure durations for toxicants.

Biotransformation Reactions

Most toxicants that enter body tissues are lipophilic. The chemical reactions responsible for changing a lipophilic toxicant into a chemical form which the body can eliminate are termed phase I and phase II biotransformations. The goal of the phase I and phase II biotransformation reactions is to facilitate detoxication (i.e., detoxification), thus producing water-soluble metabolites that are more readily eliminated by the urinary and biliary (pertaining to liver bile) systems (Figure 6-1).

At physiologic pH, a toxicant or its metabolites that are water soluble will undergo dissociation into ions or become ionized. Ionized molecules are the molecules that *react* in living systems. These ionized molecules (e.g., toxic metabolites), with their positively or negatively charged regions, are the molecules that are more readily transported across cell membranes.

On occasion, biotransformation produces intermediate or final metabolites possessing toxic properties not found in the original parent chemical, be it a naturally occurring endogenous chemical or a xenobiotic. The terms bioactivation and toxication refer to the sequence of chemical reactions that produce inter-

mediate or final metabolites that are more toxic (or reactive) than the original parent chemical. In some cases bioactivation produces a highly reactive metabolite that may interact with naturally occurring macromolecules within the cytoplasm, cell membrane, or nucleus (e.g., DNA).

Once distributed, the original chemical may not exhibit any toxic properties and should be referred to as a xenobiotic, not a toxicant. An example is acetaminophen, a commonly used analgesic (pain killer) and antipyretic (fever reducer). When prescribed doses are taken, this drug gives the desired therapeutic response with little or no resulting toxicity. This is because acetaminophen rapidly undergoes phase I and phase II biotransformation reactions and is subsequently eliminated in the urine and feces. However, at high doses, normal phase I and phase II biotransformation reactions are overwhelmed, and an additional biosynthetic pathway produces a reactive metabolite toxic to the liver (hepatotoxic).

Location of Biotransformation Reactions

Most tissues have a limited ability to biotransform. For example, skin, testes, and placenta have a low capacity, whereas the intestines, kidneys, and lungs have a medium capacity. The highest capacity for biotransformation is in the liver. The liver receives blood directly from the gastrointestinal tract, where chemicals, nutrients, and toxicants are absorbed. Blood, with its gastrointestinally derived "chemical payload," is eventually distributed to all other tissues. It is vital that the liver removes

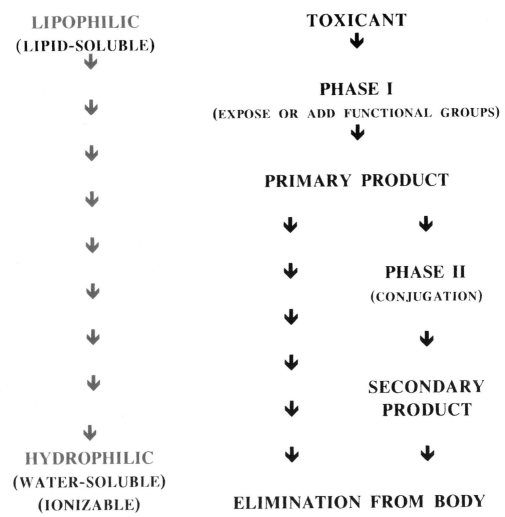

Figure 6-1. Schematic relationship among toxicants, phase I and phase II biotransformations, products, and elimination from the body in relation to lipophilic and hydrophilic characteristics.

potentially toxic chemicals from the blood prior to distribution; the "first pass" means that blood from the gastrointestinal tract is shunted directly to the liver via the portal vein, thereby increasing the possibility of immediate detoxication.

The liver's biotransformation capacity is not specific for toxicants. Rather, the liver uses phase I and phase II biotransformation reactions that, in addition to the "normal" work of biotransforming endogenous chemicals (e.g., bilirubin) and xenobiotics, are also capable of chemically modifying toxicants to facilitate their elimination from the body.

At the subcellular level the enzymes that catalyze biotransformation reactions occur either free in the cytoplasm or bound to the membrane of the endoplasmic reticulum in hepatocytes as well as other cells in the body. Microsomal enzymes are associated with phase I reactions. The term microsomal describes the "small bodies" or vesicles that form when hepatocytes or liver tissue is homogenized (blended) to form an acellular homogenate (pureed hepatocytes!). Within the homogenate, small segments of endoplasmic reticulum membranes with bound phase I enzymes spontaneously form small vesicles.

The hepatocyte homogenate also contains *soluble* enzymes that catalyze biotransformation reactions. These cytosolic enzymes are non-membrane-bound and occur free within the cytoplasm. Cytosolic enzymes are associated with phase II reactions.

Although small and often overlooked, microbes living in the intestine are capable of biotransformation. The role that the more than 400 bacterial species play in the biotransformation of xenobiotics is probably equal to that of the liver.

Factors Affecting Biotransformation

How effective biotransformation is in detoxication of toxicants absorbed and distributed in the human body depends on several factors, including age, gender, nutrition, disease, and time of day. Biotransformation characteristics associated with different organisms are unique. Differences in the qualitative and quantitative properties of phase I and phase II biotransformation enzymes must be considered. Since nonhuman species are often used for toxicity testing, caution should be exercised when extrapolating or generalizing conclusions from species other than our own.

In general, human fetuses and neonates (newborns) have limited abilities for xenobiotic biotransformations. This is due to inherent deficiencies in many, but not all, of the enzymes responsible for catalyzing phase I and phase II biotransformations. The capacity for biotransformation develops rapidly throughout infancy and peaks during adolescence and adulthood. In the aged (over age 65), the levels of many enzymes—including those related to biotransformation—have declined, predisposing them once again to the effects of toxicants.

Experimental studies on rodents (e.g., mice and rats) indicate that there are significant male and female differences in the capacity to biotransform. For some of the enzymes that catalyze phase I and phase II reactions, male rodents exhibit five times the capacity to biotransform as compared to females. Gender differences are also suspected to be present in humans and are likely related to variation in tissue mass, V_D, enzyme concentrations, hormone levels, and protein binding. Animal studies indicate that circadian (L. *circa*, about; -*dian*, a day) rhythms influence the rate of biotransformation of xenobiotics. The concentrations of many chemicals and enzymes related to biotransformation are known to fluctuate during the day.

Biotransformation is also affected by nutritional status, as evidenced in nonhuman studies. Certainly caution should be exercised when extrapolating results from animal studies to humans. It is not unreasonable to conclude that humans could exhibit many of the same responses as those observed in animals with dietary deficiencies in vitamins, minerals, and nutrients—resulting in a decline in biotransformation rates.

Many diseases impair an individual's capacity to biotransform xenobiotics, including toxicants. Hepatitis, an inflammatory liver disease, can reduce hepatic biotransformation to half of the normal capacity. The liver is a key site of biotransformation, and its impairment poses serious health consequences.

Phase I Reactions

During phase I biotransformation reactions a small polar group is either exposed ("unmasked") on the toxicant or added to the toxicant (Figure 6-2). The polar group enhances the solubility of the toxicant in water, which favors elimination. The reactions are catalyzed (brought about) by nonspecific enzyme systems, the most important of which is cytochrome P-450.

Cytochromes (*cyto-*, cell; *-chrome*, colored) are iron–protein complexes that transport electrons (or hydrogen) by changing the valency of iron (e.g., $Fe^{++} \rightarrow Fe^{+++} + e^-$ or $Fe^{+++} + e^- \rightarrow Fe^{++}$). Cytochrome P-450 gets it name from the observation that, in its reduced state (i.e., Fe^{++}), this iron–protein complex has a maximum absorbance of visible light at 450 nm (1 nanometer = 10^{-9} meters; the part of the spectrum that is visible to the human eye ranges from violet at 390 nm to red at 760 nm).

Phase I reactions usually involve: (1) oxidation, which occurs when the toxicant loses electrons, (2) reduction, when the toxicant gains electrons, or (3) hydrolysis, a process that cleaves (splits) the toxicant into two or more simpler molecules, each of which then combines with a part of water (i.e., H^+ and OH^-) at the site of cleavage.

Toxicants undergoing phase I biotransformation will result in metabolites sufficiently ionized, or hydrophilic, to be either readily eliminated from the body without further biotransformation reactions required or rendered as an intermediate metabolite ready for phase II biotransformation. Some intermediate or final metabolites may be more toxic than the parent chemical.

Phase II Reactions

On completion of a phase I reaction, the new intermediate metabolite produced contains a reactive chemical group (e.g., hydroxyl, -OH; amino, $-NH_2$; or carboxyl, -COOH). For many intermediate metabolites the reactive sites, which were either exposed or added during phase I biotransformation, do not confer sufficient hydrophilic properties to permit elimination from the body. These metabolites must undergo additional biotransformation, called a phase II reaction.

During phase II reactions (Figure 6-3) a molecule provided by the body must be added to the reactive site produced during phase I. Phase II reactions are referred to as conjugation reactions. These reactions produce a conjugate metabolite that is more water-soluble than the original toxicant or phase I metabolite. In most instances the hydrophilic phase II metabolite can be readily eliminated from the body.

One of the most popular molecules added directly to the toxicant or its phase I metabolite is glucuronic acid, a molecule derived from glucose, a common carbohydrate (sugar) that is the primary source of energy for cells.

REACTION	EXAMPLE
N-OXIDATION	$RNH_2 \rightarrow RNHOH$
S-OXIDATION	$\begin{array}{c} R_1 \\ \backslash \\ S \\ / \\ R_2 \end{array} \rightarrow \begin{array}{c} R_1 \\ \backslash \\ S{=}O \\ / \\ R_2 \end{array}$
CARBONYL REDUCTION	$\begin{array}{c} RCR' \\ \| \\ O \end{array} \rightarrow \begin{array}{c} RCHR' \\ \| \\ OH \end{array}$
HYDROLYSIS (ESTERS)	$R_1COOR_2 \rightarrow R_1COOH + R_2OH$
DESULFURATION	$\begin{array}{c} R_1 \\ \backslash \\ C{=}S \\ / \\ R_2 \end{array} \rightarrow \begin{array}{c} R_1 \\ \backslash \\ C{=}O \\ / \\ R_2 \end{array}$
DEHYDROGENATION	$RCH_2OH \rightarrow RCHO$

Figure 6-2. Representative phase I biotransformation reactions.

Glucuronidation, the process of adding glucuronide to the toxicant or phase I metabolite, occurs primarily in hepatocytes, the functional cells of the liver. The resulting glucuronic acid conjugate is excreted into the bile, which then moves on to the intestine for elimination in the feces. Typically, glucuronic acid conjugates with MW > 350 are secreted in the bile, while those with MW < 250 are secreted by the kidney.

Another phase II reaction is sulfate conjugation, which takes place primarily in the liver. Unlike glucuronic acid conjugates that are eliminated in the

REACTION	REACTANT	EXAMPLES
GLUCURONIDATION	GLUCURONIDE	PHENOL TRICHLOROETHANOL NICOTINIC ACID
METHYLATION	S-ADENOSYLMETHYL -TRANSFERASE (SAM)	CATECHOLS INDOLAMINES THIOURACIL
SULFATE CONJUGATION	PHOSPHOADENOSYL PHOSPHOSULFATE (PAPS)	PHENOL TOLUENE ACETAMINOPHEN

Figure 6-3. Representative phase II biotransformation reactions.

bile, the highly polar sulfate conjugates are readily secreted in the urine. Other phase II reactions may involve the addition of a methyl group ($-CH_3$) or an amino acid, most commonly glycine. Sodium salicylate (aspirin) is eliminated as a glycine–salicylic acid conjugate in the urine.

Elimination of Toxicants

Toxicants or their phase I or phase II metabolites are eliminated from the body by many different routes, including urine, feces, exhaled air from the lungs, milk, sweat, saliva, and cerebrospinal fluid. By far the main route of elimination is the urine, produced by the kidneys. Second is fecal elimination, which involves excretion of xenobiotics into the bile by hepatocytes. Third is elimination via the lungs, where gaseous toxicants or their metabolites are exhaled during the respiratory cycle.

Urinary Elimination

As an organ of elimination the kidney allows for intimate contact between the circulatory system and the urinary system. Two kidneys, each weighing about 150 g, are present in adults. Each kidney contains about one million nephrons—the functional units of the kidney (Figure 6-4). Three distinct structural/*functional* regions characterize each nephron: (1) Bowman's capsule/*filtration*, (2) proximal tubule/*reabsorption*, and (3) distal tubule/*secretion*. These structural regions, straightened and laid end to end, would extend over 75 miles.

Filtration is the first process in urine formation. The glomerulus, with its dense capillary network, brings blood into close contact with Bowman's capsule in the nephron. Glomerular filtration results when hydrostatic pressure within the capillaries forces small molecules, including water, across the sieve-like filter and into Bowman's capsule.

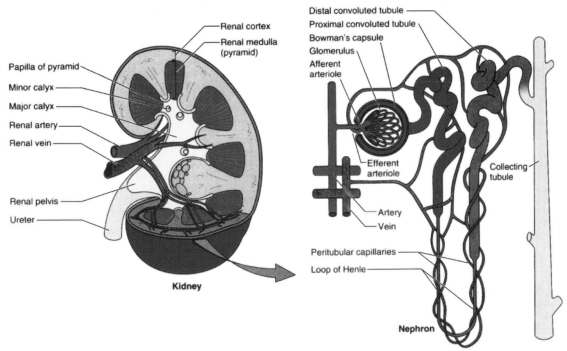

Figure 6-4. Kidney and nephron anatomy. (From M. C. Willis, *Medical Terminology: The Language of Health Care.* Williams & Wilkins, 1996. Reproduced by permission.)

Molecules with molecular weights greater than 60,000, which include large protein molecules and red blood cells, remain in the capillary and do not become part of the urinary filtrate.

In adults, the glomerular filtrate accumulates at a rate of about 125 mL/min or 180 L/day. This volume of filtrate is not entirely unexpected, considering blood flow to the kidneys is over 400 mL/100 g/min, and that the kidneys receive about one-fourth of the total cardiac output. Fortunately, processes occurring in "downstream" structures of the nephron return about 99% of the watery filtrate, along with important small molecules, to the blood supply.

Reabsorption, the second process in urine formation, occurs in the proximal convoluted tubule of the nephron. In this region most of the water lost during glomerular filtration reenters the blood. All of the glucose, potassium, and amino acids are reabsorbed, either by passive or active transport mechanisms, into the blood in the proximal convoluted tubule. Reabsorption of water occurs by osmosis, a passive transport process in which water follows its own concentration gradient and moves from a region of high concentration in the proximal tubule into a region of lower concentration in the capillaries surrounding the tubule.

Secretion, which occurs in the distal convoluted tubule, is the last process in urine formation. Whereas reabsorption is responsible for moving water and small molecules out of the urine and into the blood, secretion transports molecules out of the blood and into the urine. Secreted substances include potassium ions, hydrogen ions, and some xenobiotics.

The very processes used to eliminate endogenous metabolites are the processes used to rid the body of toxicants or toxicant metabolites entering the systemic circulation. Urinary elimination may result from glomerular filtration, or passive or active tubular transport of the toxicant into the urine. Some toxicants enter the urinary filtrate during glomerular filtration. If unaffected by an additional process, such as reabsorption, the toxicant will remain in the urine and be eliminated from the body. This is particularly true for small, polar toxicants. Other toxicants too large to enter the urinary filtrate may move via passive transport from the blood across capillary endothelial cells and nephron tubule membranes to enter the urine. Some protein-bound toxicants, which are too large to enter urine during glomerular filtration, may be transported into the urine by tubular secretion.

Fecal Elimination

Ingested xenobiotics will either be unabsorbed, and pass on through in the feces, or be absorbed, and subsequently enter the circulatory system. For many xenobiotics the former is a path that poses little hazard to the body; however, once absorbed and distributed, xenobiotics (including toxicants) will require urinary, fecal, or other routes of elimination. Not all toxicants will be eliminated from the body by the same route that they entered. Toxicants or their metabolites may enter feces as a result of intestinal or biliary excretion.

Intestinal excretion involves the transport of xenobiotics from the blood into the intestinal lumen. Recall that the submucosa of the intestines is highly vascularized and, once absorbed, nutrients and xenobiotics are rapidly distributed. During intestinal excretion the reverse occurs. Xenobiotics passively diffuse through the endothelia of capillaries in the submucosa, then through mucosal cells and into the intestinal lumen to be eliminated in feces.

Fecal excretion is an important process leading to the elimination of toxicants and their metabolites. It involves not only the gastrointestinal system but also an accessory organ, the liver. Cells in the liver (hepatocytes) secrete bile. Hepatocytes are capable of producing about 500 mL of bile per day. Biotransformed (phase I) toxicants, with possible phase II conjugates, flow into small ducts called canaliculi, which in turn flow into larger bile ducts and the gallbladder for temporary storage. To facilitate digestion, bile is released. A bile duct connects the liver to the duodenum of the small intestine.

Bile is composed of water (97%), bile salts (0.7%), inorganic salts (0.7%), bile pigments (0.2%), and trace amounts of fatty acids, fats, cholesterol, and lecithin. In addition to transporting xenobiotics or their conjugates, bile has other essential functions. For example, bile salts are responsible for the emulsification of fats in the duodenum. This facilitates fat digestion and absorption in the small intestine.

Not all toxicant- or metabolite-conjugates transported in bile from the liver to the intestines are eliminated in feces. Some are *re*absorbed as they pass through the small intestine and will enter the circulatory system for transport via the portal vein back to the liver. When phase II conjugates become hydrolyzed (the toxicant is split from the endogenous molecule) the "free" toxicant, usually lipophilic, can be reabsorbed through the intestinal mucosa. This cycling of toxicants between the liver and intestine is termed an entero-hepatic loop. Bile salts participate in their own enterohepatic loop—over 90% are reabsorbed in the intestines for a return trip to the liver.

Nutrient—and potentially toxicant—rich blood leaves the intestines and circulates through the liver (1.5 L/min) before flowing on to the heart and lungs, after which it is pumped into the systemic circulation for distribution to tissues. The capacity to biotransform toxic substances before they can be distributed to other body regions is of survival value.

Respiratory Elimination

The lungs are involved in the elimination of xenobiotics that exist in a gaseous phase at body temperature. Simple diffusion results when the concentration (vapor pressure) of the xenobiotic dissolved in capillary blood is greater than the gaseous phase present in the alveoli. This represents a concentration gradient, and the volatile gas will diffuse down its concentration gradient until equilibrium across the alveolar membrane is achieved. Gases with a low solubility in blood are more rapidly eliminated than those gases with a high solubility. In addition to vapor pressure and solubility, other factors such as respiration rate and blood flow to the lungs determine the elimination of gaseous xenobiotics by the lungs.

Ethanol, a volatile alcohol, provides a good example of a xenobiotic that is eliminated from the body by fecal, urinary, and respiratory routes. After the ingestion of alcoholic beverages, ethanol is rapidly absorbed across the gastrointestinal mucosa. Ethanol then enters the systemic circulation and is distributed to the tissues, including the brain, kidneys, liver, and lungs. About 90% of the ethanol undergoes phase I biotransformation to acetaldehyde and acetate in the liver. The remaining 10% is eliminated unchanged by the kidneys in urine or by the lungs during exhalation.

Additional Routes of Elimination

Saliva. Three pairs of salivary glands located within the oral cavity are capable of producing about 1.5 L of saliva per day. Xenobiotics that passively diffuse into saliva may, when swallowed, be available for absorption through the mucosa of the gastrointestinal system. Overall, saliva plays a minor role in the elimination of toxicants. However, the elimination of some pharmaceuticals (drugs) into saliva is responsible for the "drug taste" reported by some patients.

Sweat. Skin has about 80 sweat glands/cm^2. The amount of water lost each day in sweat varies widely and may amount to more than 1 L/hour under strenuous work in hot temperatures. On average, about 100 mL of water is lost per day. This represents about 4% of water output from all sources (2,400 mL), such as water expired in air (350 mL), water lost through skin by diffusion (350 mL) and sweat (100 mL), and water lost in urine (1,400 mL) and intestines (200 mL). Sweat is responsible for the elimination of many metals, including cadmium, copper, iron, lead, nickel, and zinc. Under normal conditions xenobiotics entering sweat through passive diffusion represent a minor elimination component; however, under conditions of greater sweat production, significant elimination may result.

Milk. During lactation, xenobiotics present in the mother's blood may also be detectable in her milk. This is partly due to the fat content of milk, which enhances the passage of lipophilic xenobiotics. The concentration of a given xenobiotic eliminated in milk depends on the chemical characteristics of the xenobiotic, blood flow to breasts, and amount of milk produced. (Note: Breastfeeding should be discouraged when toxicants are present in the mother's milk, as the toxicants will likely have an adverse effect on the infant.)

Nails. Fingernails and toenails are part of the integumentary system and are shed under normal growth conditions. As the nail forms, xenobiotics (e.g., arsenic) may become incorporated into the horny matrix. This effectively removes the xenobiotic from the body as the nail is worn away.

Hair. Like nails, hair is a part of the integumentary system and under normal growth conditions it is, over time, lost from the body. Xenobiotics (such as arsenic, cadmium, and lead) that become incorporated in hair will eventually be eliminated from the body.

Skin. Desquamation, or loss of epithelial cells in the epidermis, permits the loss of xenobiotics through the skin.

Cerebrospinal Fluid. Toxicants that enter the central nervous system (CNS includes the brain and spinal cord) may enter the cerebrospinal fluid (CSF). About 150 mL of CSF is present in the ventricles of the brain and in the tissues (i.e., subarachnoid space) that surround the brain and spinal cord. CSF is produced mostly by the choroid plexus at a rate of 550 mL/day. The CSF turnover rate is about 3.7 times/day, which contributes to a substantial flow of this fluid. CSF is absorbed into the tissues surrounding the CNS where it enters the venous blood supply. Toxicants in the CSF are actively transported into tissues that surround the brain or by passive diffusion through the blood-brain barrier.

Review Questions

1. Which is *not* a true statement about metabolites that result from biotransformation?

A. They are chemically distinct from the original toxicant.
B. They have a reduced ability to cross membranes.
C. They are usually more hydrophobic than the original toxicant.
D. A and B
E. A, B, and C

2. If exposure to a toxicant resulted in blood plasma concentration of 200 mg/mL, and the toxicant has a $T_{1/2}$ of 4 hours, then how long would it take to reduce the blood plasma level of the toxicant to 25 mg/mL?

A. 8 hours
B. 12 hours
C. 16 hours
D. 24 hours
E. More than 25 hours

3. Term used to describe the sequence of biotransformation reactions that produce intermediate or final metabolites that are more toxic than the original parent chemical.

A. Bioactivation
B. Detoxication
C. Ionization
D. Oxidation
E. Reduction

4. Which sequence best represents the *highest to lowest* capacity for biotransformation to take place in different organs?

A. Skin→lungs→liver
B. Lungs→liver→kidneys
C. Placenta→testes→lungs
D. Liver→kidneys→skin
E. Liver→skin→lungs

5. Which is *not* a true statement about phase I biotransformations?

A. They involve cytosolic enzymes.
B. They may be impaired by disease states.
C. Nonspecific reactions are catalyzed by cytochrome P-450.
D. They produce metabolites sufficiently ionized to be eliminated from the body.
E. They may involve oxidation, reduction, or hydrolysis reactions.

6. Some protein-bound toxicants, which are too large to enter urine during glomerular filtration, may be transported into the urine by tubular secretion.

A. True
B. False

7. Which is (are) a true statement about fecal elimination of toxicants?

A. Unabsorbed toxicants will pass on through in the feces.
B. Intestinal excretion involves the transport of toxicants from the blood into the intestinal lumen.
C. Fecal excretion is the most important process leading to the elimination of toxicants and their metabolites.
D. A and C
E. A, B, and C

8.   The lungs are involved in the elimination of toxicants that exist in a gaseous phase at body temperature.

A.   True
B.   False

9.   Discuss additional routes of elimination other than urinary, fecal, and respiratory pathways.

10.   How are phase I and phase II biotransformation reactions influenced by the following: age, gender, nutrition, disease, and species?

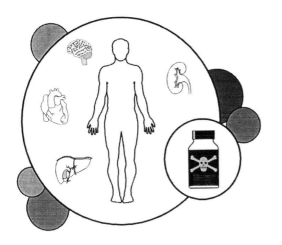

O bjectives

- Define target organ toxicity

- Explain the basis for the specificity of organ toxicity

- Contrast the toxicity mechanisms for hematotoxicity, hepatotoxicity, nephrotoxicity, neurotoxicity, dermatotoxicity, and pulmonotoxicity

- Describe examples of target organ toxicity

- Discuss the characteristic evaluative procedures for determining toxicity in target organs

K eywords

acute tubular necrosis (ATN)
agranulocytopenia
allergic contact dermatitis
anemia
anthracosilicosis
anuria
arterial blood gases (ABGs)
asbestosis
berylliosis
blood dyscrasias
blood urea nitrogen (BUN)
bronchoscopy
central nervous system (CNS)
chloracne
creatinine
dermatotoxicity
erythropoietin
forced vital capacity (FVC)
glial cells
glomerular filtration rate (GFR)
glycosuria
hematotoxicity
hematotoxins
hematuria
hemolytic anemias
hepatotoxicity
hypoxia

eywords *(continued)*

inulin
irritant contact dermatitis
leukemia
microcytic hypochromic anemia
myelin
nephritic syndrome
nephrotic syndrome
nephrotoxicity
nerve
neurons
neurotoxicity
neurotransmitter
obstructive uropathies
oliguria

PAH clearance
pancytopenia
peripheral nervous system (PNS)
phototoxicity
pneumoconioses
poietins
proteinuria
pulmonary fibrosis
pulmonotoxicity
radiopharmaceutical
silicosis
slow vital capacity (SVC)
target organ toxicity
thrombocytopenia

Introduction to Target Organ Toxicity

Time-dependent toxicokinetic processes related to absorption, distribution, storage, biotransformation, and elimination will determine how much of a toxicant will be distributed to a specific target organ (e.g., kidney, liver, or lung) in the body. Target organ toxicity is defined as the adverse effects or disease states manifested in specific organs in the body. Toxicity is unique for each organ, since each organ is a unique assemblage of tissues, and each tissue is a unique assemblage of cells. Toxicity may be enhanced by distribution features that deliver a high concentration of the toxicant to a specific organ (e.g., large blood flow/mass ratio) or by inherent features of the cells and tissues of the organ that render it highly susceptible to the toxicant, even at low concentrations.

Although the mechanisms responsible for organ toxicity are not always known, the observed differences in target organ toxicity are most likely due to structural and functional differences in the cells that make up the tissues and organs. In other words, under the influence of a toxicant, each organ will manifest different disease states (i.e., toxicity), depending on the structural and functional characteristics of the cells present.

Cells differ in many ways, including their energy consumption (e.g., use of ATP), rate of cellular division, active and passive transport characteristics, relationship to cell barriers (e.g., blood-brain barrier) and extracellular matrices (e.g., hydroxyapatite, collagen), presence of intracellular components (e.g., contractile filaments in muscle cells, microtubules in neurons), repair mechanisms, and biotransformation capacity. Cellular specialization, in addition to making "life" possible in a multicellular organism, also means that each organ will respond to a toxicant in a different way.

When considering target organ toxicity, remember that: (1) not all organs are affected to the same extent by a toxicant; (2) a single toxicant may have several target organs; (3) several toxicants may have the same target organ; (4) the highest concentration of a toxicant is not always found in the target organ; and (5) the concentration of a toxicant in a target organ is the result of *all* toxicokinetic processes.

Hematotoxicity

Introduction

Although not self-evident, blood is classified as a connective tissue. All blood cell components (i.e., erythrocytes, leukocytes, and thrombocytes) have their origin in stem cells found in bone marrow (Figure 7-1). Poietins, or stimulating factors, regulate the "fate" of a stem cell—that is, whether it becomes an erythrocyte (erythropoietin), leukocyte, or thrombocyte. Proper oxygen transport, immune function, and clot formation result when normal numbers of erythrocytes, leukocytes, and thrombocytes are present, respectively.

Toxicity Mechanisms

Hematotoxins (G. *haimatos*, blood) alter quantitative and qualitative characteristics of blood cells to produce

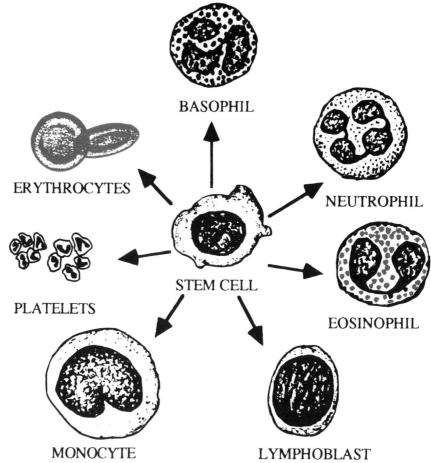

BASOPHIL

NEUTROPHIL

ERYTHROCYTES

STEM CELL

EOSINOPHIL

PLATELETS

MONOCYTE

LYMPHOBLAST

Figure 7-1. Formation of the cellular components of blood from stem cells.

toxic symptoms. Hematotoxicity occurs when too many or too few of these different blood components are present or structural anomalies occurring in blood components interfere with normal functioning.

For example, anemia occurs when there are too few erythrocytes, which results in a reduced oxygen-carrying capacity of the blood. Too many white blood cells, leukemia, can be deadly. Not enough thrombocytes, thrombocytopenia, may result in external or internal hemorrhaging (i.e., blood loss).

Qualitative changes in blood cell components can also result in disease.

Microcytic hypochromic anemia results when erythrocytes have a low hemoglobin content. Although adequate in quantity, the quality of these "small, less colored" red blood cells prevents them from carrying the normal amount of oxygen.

Examples

Carbon monoxide (CO) alters the oxygen-carrying capacity of hemoglobin. The *qualitative* change occurs because CO preferentially binds to hemoglobin, preventing the transport of O_2. In effect, CO out-competes O_2 for the available transport sites on hemoglobin molecules, leading to hypoxia or anoxia, which is an inadequate supply of oxygen to the tissues. Symptoms may include low blood pressure, fainting, dizziness, headache, weakness, and nausea.

Cyanide (HCN) and hydrogen sulfide (H_2S) are capable of producing cytotoxic hypoxia, a potentially lethal condition in which cells in the body cannot utilize O_2 during normal cell metabolism associated with energy production. Oxygen in the venous blood becomes abnormally high, since it is not being used by the cell. Hydrogen sulfide, as encountered in oil fields and sewers, is recognized by its "rotten egg" odor.

Blood dyscrasias (i.e., disorders) can be induced by toxicants. Dyscrasias usually result in abnormal cellular components—too many of one blood cell type, too few of another. Benzene (an organic solvent) is known to cause thrombocytopenia and leukemia. Agranulocytopenia, a decrease in the monocytes and lymphocytes (i.e., agranular leukocytes), can be induced by DDT (an insecticide). Other chemicals, such as chlordane, can cause pancytopenia, a reduction in all blood cells. Finally, hemolytic anemias result when erythrocytes are destroyed by a toxicant, such as naphthalene.

Evaluating Hematotoxicity

Five tests are commonly used to measure the quantitative and qualitative aspects of blood. Three of these—the red cell count, white cell count, and platelet count—are quantitative measurements of blood components. The other two, hemoglobin (Hb) and hematocrit (HCT), are indicators of the oxygen-carrying capacity of blood. Collectively, these tests, plus others, are termed a complete blood count (CBC) with differential (DIFF). The CBC/DIFF provides information on the number, variety, percentages, concentrations, and quality of blood components. Additionally, arterial blood gases (ABGs) may be measured to determine the partial pressure of oxygen (PaO_2), partial pressure of carbon dioxide ($PaCO_2$), and the percentage of hemoglobin bound to oxygen or oxygen saturation (SaO_2), all of which are important indicators of the blood's ability to acquire and release oxygen to the tissues.

Hepatotoxicity

Introduction

Hepatotoxicity refers to toxic effects in the liver. In addition to its important role in detoxication, this largest gland in the body also functions to synthesize many proteins (e.g., plasma albumin, coagulation proteins), excrete bile, and metabolize fats, carbohydrates, and proteins.

The liver is particularly susceptible to toxic agents for two reasons: (1) after absorption, most toxicants that enter the blood flow through the liver ("first pass") before being distributed to other systemic regions, including other organs; and (2) the liver is the primary site for biotransformation of toxicants, which exposes the liver to toxicants and their metabolites, some of which, as a result of toxication or bioactivation, are more toxic than the original chemical.

The functional unit of the liver is the lobule. Approximately 50,000–100,000 lobules are present in an adult liver. Each lobule is about 1–2 mm in diameter and contains an orderly arrangement of hepatocytes, a hepatic venule, a hepatic arteriole, sinusoids, and a small bile duct called a canaliculus (Figure 7-2). As blood flows through sinusoids in the lobule, from hepatic arterioles and portal venules to the terminal hepatic venule, bile is formed by hepatocytes that lie between vascular sinusoids and the canaliculi. Bile flows within canaliculi in the opposite direction of the blood flow.

Toxicity Mechanisms

Liver toxicants are typically characterized as being cytotoxic or cholestatic (*chole-*, bile; *-static*, standing still). Cytotoxic mechanisms affect hepatocytes and are responsible for different types of liver injury, including fatty liver, liver necrosis, and cirrhosis. Cholestatic mechanisms affect the flow of bile. *Intrahepatic* cholestasis occurs when the flow of bile is blocked *within the liver* as it flows through canaliculi, as well as bile ductules.

Examples

A number of organic chemicals are toxic to hepatocytes, including trichloroethylene, carbon tetrachloride, and dichlorodiphenyltrichloroethane (DDT). The resulting toxicity may lead to necrosis (the death of hepatocytes), inflammation, or cirrhosis (the replacement of hepatocytes by fibrous tissue during the repair of damaged hepatocytes). Chronic ethanol toxicity involves the accumulation of excess fat within hepatocytes, which can lead to fatty liver disease and cirrhosis.

The cholestatic mechanisms that lead to the blockage of bile are poorly understood. "Blockage" may result from blocked transport mechanisms in the cell membrane of hepatocytes, in which case little or no bile is transported into the canaliculi, or from precipitates and "bile plugs" that form within canaliculi. Bile salts, steroids, and α-naphthylisocyanate are known cholestatic agents.

Evaluating Hepatotoxicity

Noninvasive liver tests are used to evaluate liver structure and function. Typically three types of tests are done. The first involves testing blood serum (i.e., the noncellular portion) for particular enzymes known to be present at specific levels when the liver is functioning properly, but that may be elevated in a damaged liver due to their release into the blood from damaged hepatocytes.

The second type of test, again performed on blood, examines liver function for its ability to remove routinely encountered substances (e.g., bilirubin) or introduced substances (e.g., dyes) from the blood. Since the liver is responsible for producing most of the factors involved in the cascading sequence of reactions needed to produce a "fibrin clot," the time for clot formation can be examined as an indicator of liver function.

A third type of test, the liver scan, is used to examine both liver anatomy (e.g., size) and function (e.g., flow of blood and bile). This nuclear test is performed by intravenously injecting a radiopharmaceutical, a radioactive chemical used in diagnostics. Pictures using noninvasive radioactive imaging tech-

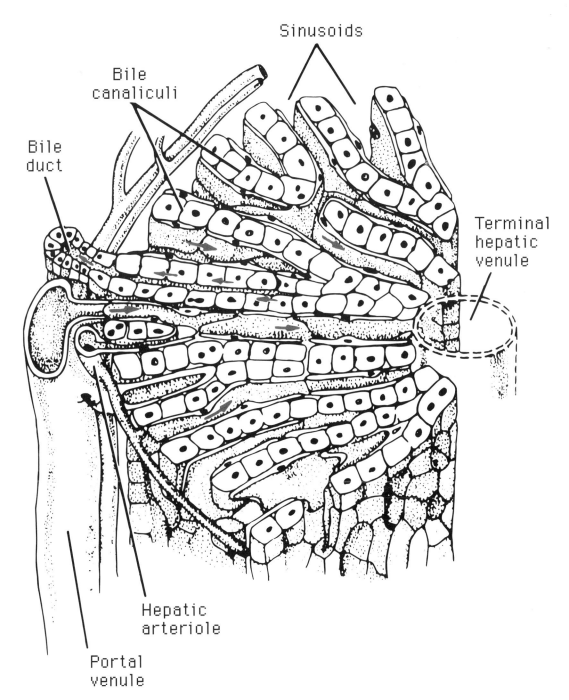

Figure 7-2. Structure of a liver lobule. (From D. W. Fawcett, *Bloom and Fawcett, A Textbook of Histology*, 11th edition. W. B. Saunders, 1986. Reprinted by permission.)

niques are then taken to detect the radio-pharmaceutical as it circulates through the liver.

Nephrotoxicity

Introduction

Nephrotoxicity refers to the toxic effects in the kidney. Remember, three processes occurring in the kidney, glomerular filtration, tubular reabsorption, and tubular secretion, are responsible for the production of urine. Nephrotoxins are known to influence each of these processes.

Toxicity Mechanisms

During filtration, two toxic responses may be manifest. First, due to the large volume of filtrate formed at the interface of the vascular glomerulus and Bowman's capsule, toxicants may accumulate in this anatomical region of the nephron. Second, this may in turn increase or decrease the rate of filtration or alter characteristics of the glomerular apparatus (the filter). If the glomerulus is made more porous, or less selective, substances normally excluded by the filter will be able to cross and enter the filtrate.

Other nephrotoxicity mechanisms affect both the qualitative and quantitative aspects of the reabsorptive process. The proximal tubule is responsible for the selective reabsorption of most of the salts and water present in the filtrate. All amino acids and glucose are reabsorbed and returned to the blood in this region. Any changes induced by toxicants or their metabolites in the characteristics of the cell membranes that form

the tubule can profoundly affect the reabsorptive process.

Tubular secretion, the third and last process of urine formation, is responsible for the active transport of substances from the blood (e.g., H^+, K^+, xenobiotics) into the urine. Again, when toxicants or their metabolites alter these transport mechanisms, the nephron is unable to function properly and nephrotoxicity may result.

Examples

The pathologies associated with nephrotoxicity are dependent on the anatomical region of the nephron affected by the toxicant. Two major responses may be observed when the glomerular filtration apparatus is injured: nephrotic syndrome and nephritic syndrome. Although the pathologies for each are complex, nephrotic syndrome is usually characterized by heavy proteinuria (i.e., presence of protein in the urine), whereas nephritic syndrome is typically characterized by hematuria (i.e., presence of blood cells in the urine). Some xenobiotics, such as lead and heroin, are linked to nephrotic syndrome, often resulting in heavy proteinuria.

The selectivity of the glomerular filter can be altered by exposure to xenobiotics. In contrast to *increased* permeability to albumin resulting from exposure to puromycin (an antibiotic), two other antibiotics—gentamycin and kanamycin—*decrease* the rate of glomerular filtration.

As the glomerular filtrate flows through the nephron, renal tubules may be exposed to high concentrations of filtered toxicants or their metabolites. Damage to the epithelial cells that line the tubules is responsible for producing acute tubular necrosis (ATN). Heavy metals, antibiotics, and organic solvents are known to cause ATN.

Tubular reabsorption is affected by cadmium (Cd), lead (Pb), and mercury (Hg). Up to the age of 50, Cd normally accumulates in the human kidney—in fact, about 10 times as much Cd accumulates in the kidney as in the liver. Cadmium is capable of producing glycosuria (loss of glucose in the urine) and aminoaciduria (loss of amino acids in the urine). Lead inhibits the reabsorption of glucose and amino acids in the proximal tubule, also leading to glycosuria and aminoaciduria. Kidney failure may result from exposure to inorganic mercury (Hg^{2+}), a powerful tubular nephrotoxin. When oliguria (little urine) or anuria (no urine) is formed, the buildup of toxic wastes in the body can lead to death.

Obstructive uropathies result when the flow of urine is prevented either by intratubular or extratubular pathologies. Ethylene glycol, a commonly used antifreeze or coolant, is metabolized by the body to calcium oxalate. This insoluble salt accumulates in the proximal tubule, both in the lumen and in the epithelial cells lining the lumen, forming an intratubular obstruction to the normal flow of urine.

Evaluating Nephrotoxicity

A number of quantitative and qualitative tests are used to evaluate kidney function. Glomerular filtration rate (GFR) is defined as the amount of glomerular filtrate (mL) per unit of time (min). GFR can be determined by intravenously administering inulin, a fructose polymer (MW = 5,200). This sugar polymer is readily distributed in the blood,

does not become bound to plasma proteins, is not metabolized or stored, effortlessly enters the glomerular filtrate, and is not reabsorbed or secreted by the nephron. On entering the glomerular filtrate, inulin becomes a permanent component of urine. By measuring the amount of inulin present in plasma (P_I) and urine (U_I), and urine volume (V) after a specific interval of time, the inulin clearance (C_I) can be used to estimate GFR by the following formula:

$$GFR \approx (U_I)\,(V)\,/\,P_I = C_I$$

Typical values have been inserted into this formula to show its usefulness in determining normal kidney function as related to glomerular filtration rates:

$$GFR \approx (30\ mg/mL)\,(1.25\ mL/min)/$$
$$0.3\ mg/mL = 125\ mL/min$$

When nephrotoxicity affects the glomerular filtration apparatus, as may be evidenced by a decrease in the normal GFR of 125 mL/min, secondary consequences affecting toxicokinetics are likely to occur. For example, a decrease in GFR may result in a decrease in urinary elimination or "clearance" of a toxicant or its metabolites. This would increase the $T_{1/2}$ and could lead to toxicity involving other organs.

The organic acid PAH (*p*-aminohippuric acid) is used to evaluate kidney function. Readily filtered, PAH is actively secreted into the urine. It is almost completely "cleared" from blood plasma during one pass through the kidneys—so much so, that PAH clearance is used to estimate the rate of plasma flow through the kidneys. Together, inulin and PAH studies provide information about glomerular filtration and tubular secretion.

Two other indicators of kidney function are blood urea nitrogen (BUN) and creatinine. Urea (a potential endogenous toxicant formed from the catabolism of proteins) and creatinine (a product of muscle metabolism) are distributed in the blood. In adults, normally functioning kidneys will eliminate urea (25 g/day) and creatinine (1.8 g/day) into urine. Elevated serum BUN and creatinine levels are indicative of kidney dysfunction.

Neurotoxicity

Introduction

The nervous system is divided into the central nervous system (CNS) and peripheral nervous system (PNS). The CNS includes the brain and spinal cord, which function to process information and provide memory. The PNS contains peripheral nerves that are involved with sensory and motor control functions.

Neurons, more than a billion of them, are the characteristic cells found in both the CNS and PNS. These cells gather information (i.e., sensory), process information, provide memory, and initiate appropriate responses (i.e., motor). Neurons are composed of three cellular regions: the cell body, dendrites, and axon. A nerve in the PNS is a collection of individual motor or sensory neurons.

Additional support cells in the CNS, termed glial cells, provide a structural framework and transport of nutrients (astrocytes), myelin production (oligodendrocytes), and immune function (microglia). The oligodendrocytes (CNS) and Schwann cells (PNS) are responsible for the production of myelin, a lipid-rich "cell membrane wrapping" around axons. Myelin functions to insulate the axon, enhancing the velocity of axonal conduction.

Neurons, either sensory or motor, are linked together to form a communication network. Two different processes are responsible for the propagation of a communication along a neuronal path. The first process ensures that a signal is "electrically" transmitted along the length of each neuron's axon, while the second "chemically" propagates the signal from one neuron to the next.

Neurotoxins are known to alter neurons in both the CNS and PNS, as well as the glial support cells in the CNS. Much of our current understanding of the nervous system's structure and function has resulted from the experimental use of neurotoxins—particularly neurotoxins affecting cell membrane proteins that function in cell transport and as membrane receptors.

Toxicity Mechanisms

Neurotoxins interfere with the communication ability of neurons, impeding receptor or motor neuron signaling and CNS functioning. Neurons are able to propagate a signal due to an electrical gradient that exists between the inside and outside of the axonal cell membrane. The gradient is created by Na^+/K^+ pumps in the cell membrane. These pumps actively transport Na^+ to the outside of the cell and K^+ to the inside of the cell, creating a potential difference of about -70 mV (millivolts) across the axonal cell membrane.

Within a single neuron, signal propagation occurs when the electrical wave of depolarization runs the entire length of the axon. Depolarization results when passive transport channels open to permit Na^+ to rapidly enter the cell membrane, followed almost instantaneously by additional passive channels that open to allow K^+ to exit. When passive or active transport mechanisms are slowed, blocked, or otherwise impaired, or if the membrane "leaks" excessively, the potential difference cannot be maintained and the neuron will be unable to propagate a signal down the length of the axon.

When signals reach the distal axonal region, a second process called synaptic transmission is initiated. Synaptic transmission is responsible for propagation of the signal to the next neuron. In this process a neurotransmitter (chemical messenger), is released from the distal region (i.e., presynaptic neuron). Once released (i.e., exocytosis of vesicles), the neurotransmitter moves across the synaptic cleft (a microscopically small space) to bind to receptors on the membrane surface of the postsynaptic neuron.

If sufficient neurotransmitters bind to the postsynaptic membrane receptors, the postsynaptic neuron will depolarize and the electrical wave will be propagated down the axon of the next neuron. Synaptic transmission involves a chemical mode of transmission rather than the electrical gradient responsible for propagating the signal in the axon.

Neurotoxins may bind to postsynaptic receptors, blocking synaptic transmission. Following synaptic transmission, neurotoxins may act to inhibit the "removal," or degradation, of the recently released neurotransmitters, permitting continued synaptic transmission that may be sufficient to repeatedly depolarize the postsynaptic neuron.

Examples

Neurotoxicity results when toxicants interrupt the normal mechanisms of neuronal communication. Batrachotoxin, secreted by frogs, destroys the electrical gradient by increasing the Na^+ permeability of the axonal cell membrane. Ethanol depresses CNS function by (1) inhibiting or stimulating a variety of transport channels and (2) increasing membrane fluidity by altering the packing of molecules within the membrane. Both processes act to depolarize the neuron, thereby decreasing signal transmission.

The insecticide DDT increases membrane permeability to Na^+ in the presynaptic region, which leads to continual signaling. Other insecticides (e.g., malathion, diazinon) prevent the breakdown of acetylcholine, a neurotransmitter, by binding to cholinesterase, an enzyme responsible for catabolism.

Tetrodotoxin, found in puffer fish, blocks Na^+ channels. Botulin toxin, a bacteriotoxin, prevents the release of the neurotransmitter acetylcholine. Curare, a phytotoxin, is a neuromuscular blocking agent that prevents a motor neuron from signaling a muscle cell by blocking acetylcholine receptors on muscle cells.

A number of neurotoxins (e.g., hexachlorophene and lead) damage the myelin sheath. This loss of insulation results in signal "short circuiting" between adjacent neurons and slower neuronal transmission velocities.

Evaluating Neurotoxicity

The CNS presents a challenge to direct evaluation, since invasive diagnostic procedures involving the brain and spinal cord may be life threatening. Indirect examinations provide information about neurological function, and the following are helpful in CNS and PNS toxicity diagnoses: (1) patient history, (2) mental status (e.g., memory), (3) deficits in cranial nerve function (e.g., auditory nerve function in hearing and balance), (4) sensory neuron function (e.g., pain, temperature), and (5) motor neuron function (e.g., coordination, gait, muscle strength, reflexes, tremors).

Other information related to neurological structure and function may be obtained from x rays, computerized axial tomography (CAT scans), magnetic resonance imaging (MRI), electromyography (EMG), electroencephalography (EEG), peripheral nerve conduction velocity, and cerebrospinal fluid (CSF).

Dermatotoxicity

Introduction

Dermatotoxicity describes the adverse effects produced by toxicants in the skin. Recall that skin is composed of the epidermis, dermis, and subcutaneous fatty tissue. Hair follicles, oil and sweat glands, blood vessels, and sensory neurons are present. Skin is more than a protective covering. It functions to limit water loss, reduce the harmful effects of ultraviolet radiation, prevent the entrance of microorganisms, regulate body temperature, and also biotransform and eliminate toxicants. Skin, with all its related functions, is vulnerable to toxicity, because skin is often the first part of the body to come into contact with a toxicant during handling.

Toxicity Mechanisms

Toxic skin reactions are diverse and may involve any one or a number of combinations of skin components. Irritant contact dermatitis results when toxicants elicit an inflammatory response in skin. Depending on the exposure site, this form of dermatotoxicity is manifest in the accumulation of watery fluid (edema), an increase in the amount of blood (hyperemia), or, if serious, the loss of tissue (ulceration and necrosis). Although irritant contact dermatitis is usually confined to the site of contact, prolonged exposure may result in systemic toxicity.

Some individuals show little or no response on exposure to a chemical; however, on exposure at a later time they may exhibit a delayed hypersensitivity reaction, sometimes severe, termed allergic contact dermatitis. This delayed sensitivity to toxicants involves the immune system, and it may take from a few days to many years for individuals to become sensitized. Specifically, the initial exposure sensitizes the immune system (T lymphocytes) to "recognize" the toxicant on later exposure. Again, the delayed dermatotoxicity or hypersensitivity reaction may result in edema, hyperemia, or ulceration.

Phototoxicity, a form of light-induced dermatotoxicity, results when skin is overexposed to ultraviolet light or from the combination of exposure to specific wavelengths of light *and* a phototoxic substance. Symptoms associated with phototoxicity include erythema (sunburn), hyperpigmentation, premature skin aging, and cancer. Hyperpigmentation and hypopigmentation symptoms result from changes in melanocytes, the cells located in the epidermis responsible for the production of melanin (a pigment that gives color to skin and hair).

Dermatotoxic responses may occur in hair, and in sebaceous and sweat glands. The cells responsible for hair production have some of the highest mitotic (cellular division) rates in the body. Toxicants, including chemotherapeutic agents, that interrupt cellular division typically will induce hair loss. Some dermatotoxins stimulate a proliferation of the epithelium surrounding sebaceous glands. Proliferating epithelial cells plug the pilosebaceous (L. *pilus*, hair; L. *sebaceus*, oily) orifices, producing chloracne, a condition similar to acne vulgaris experienced by adolescents.

Examples

Irritant contact dermatitis (e.g., edema, erythema) may result from exposure to a variety of agents, including organic solvents, acids (pH less than 5.5), bases, plants (orange peel, nettles), detergents, and even water. Chronic exposure to cement and chrome can result in serious ulcerative conditions accompanied by necrosis.

Allergic contact dermatitis has been linked to exposure to antibiotics (e.g., penicillin), antihistamines (e.g., diphenhydramine), anesthetics, plants (e.g., poison ivy), tanning agents, metal compounds (e.g., chromates), numerous industrial agents, and rubber antioxidants. The latter are commonly used in the manufacture of gloves, shoes, and undergarments.

Phototoxicity can result from acute and chronic exposure to ultraviolet radiation (e.g., "sunbathing"). Exposure to certain wavelengths of light coupled with simultaneous exposure to phototoxicants, such as drugs (e.g., tetracycline), perfumes, polycyclic aromatic hydrocarbons, and dyes, can result in dermatotoxicity.

Sunlight, coal tar compounds, petroleum oils, and heavy metals (e.g., arsenic, mercury) are known to produce hyperpigmentation. Of therapeutic interest in the treatment of hypopigmentation are the psoralens (e.g., 8-methoxypsoralen). Vitiligo (white patches of skin caused by an absence or decrease in melanin production) is often successfully treated with PUVA (psoralen and ultraviolet-A) therapy. Together the phototoxicant and UV-A produce hyperpigmentation.

A decrease in melanin production, or hypopigmentation, can be produced as a result of burns, chronic dermatitis, and dermatotoxicants such as *p*-tertiary butyl phenol. The mechanism by which *p*-tertiary butyl phenol functions as a depigmenting agent (i.e., melanotoxicant) is probably related to its structural likeness to L-tyrosine, one of two amino acid precursor molecules involved in melanin biosynthesis. The resulting decrease in melanin is likely due to the biosynthesis of a nonfunctional "melanin-like" molecule, or as a result of a reduction of the enzymes that normally catalyze the reaction of L-tyrosine into melanin. The enzymes are depleted or inhibited during attempts to catalyze *p*-tertiary butyl phenol, and hence are unavailable to catalyze the normal L-tyrosine to melanin reaction (Figure 7-3).

Acne-like conditions may be produced by a number of dermatotoxicants, including coal tar, greases, oils, and even cosmetics. However, a few specific halogenated aromatic com-

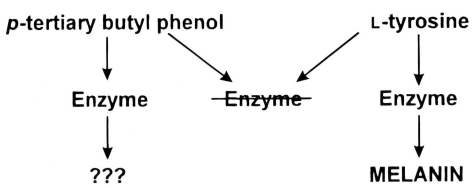

Figure 7-3. The substitution of *p*-tertiary butyl phenol for L-tyrosine in the biosynthetic pathway of melanin.

pounds are responsible for producing chloracne (Figure 7-4). Among the more active chloracne-producing chemicals are polyhalogenated biphenyls, dibenzofurans, dioxins, and naphthalenes. These halogenated chemicals induce epithelial hyperplasia (i.e., increase the number of epithelial cells), which blocks the opening to sebaceous glands.

Evaluating Dermatotoxicity

Usually the causative agents responsible for most irritant dermatitis are readily identified on questioning the patient or reviewing the history of exposure. When questioning is not helpful in isolating the causative agent, the patch test (performed by a dermatologist) may be used in diagnosis. In this test the dermatologist selects the contact allergens and concentrations to be tested. The suspected allergen is applied to the skin under a nonabsorbent patch and is left for a specified period of time. The presence of erythema and other pathologies, on removal of the patch, signifies a positive patch test result.

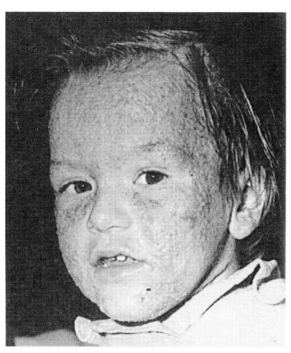

Figure 7-4. Chloracne, as seen on the face of 4-year-old Alice Senno (October 29, 1976), resulted from exposure to dioxin released when a chemical factory exploded in Seveso, Italy, on July 10, 1976. (From AP Wirephoto. Reproduced by permission.)

Pulmonotoxicity

Introduction

Pulmonotoxicity refers to the disease states in the respiratory system brought about by the inhalation of gases, vapors, liquid droplets, and particulates. Inhalational toxicants may affect nasopharyngeal, tracheobronchial, and alveolar regions.

The respiratory mucosal lining is highly susceptible to toxic substances. Unlike the stratified keratinized epithelium present in skin, the epithelial cells lining the respiratory system are not always stratified and are not keratinized, diminishing their barrier qualities. Different types of epithelial cells are present in different regions, such as the stratified squamous epithelium in the pharynx and the ciliated columnar epithelium in the tracheobronchial region. Still different epithelial cells, type I and type II pneumocytes, make up the alveoli.

The varied pulmonotoxicities are a reflection of the characteristic assemblage of the more than 40 cell types present in the respiratory system. Ultimately, concern is paramount when toxic responses decrease the lung's ability to exchange oxygen and carbon dioxide across the alveolar-capillary interface.

Toxicity Mechanisms

In addition to the systemic toxicity that may result when toxicants are absorbed through the respiratory system, the respiratory system itself exhibits responses ranging from minor local irritations and allergic responses to cellular damage, fibrosis, and neoplasms (cancer). Many gases are known to irritate the epithelial cells lining the respiratory system. Irritation may involve inflammatory responses that lead to a contraction of the smooth musculature that surrounds the air passageways (e.g., bronchioles) and to edema. These two conditions reduce the cross-sectional area of the passageway, limiting the flow of air.

Some pulmonotoxicants "target" specific cells in the respiratory system, such as the ciliated columnar epithelial cells in the tracheobronchial region or Clara cells in the region of the terminal bronchioles. The resulting pulmonotoxicity reflects the damage to these specific cell populations, which may lead to impaired function of the mucociliary escalator and necrosis, respectively—both of which decrease pulmonary function.

Asthma refers to a narrowing of the air passages in response to a number of stimuli, including allergens, infections, exercise, and drugs. Of interest to toxicologists is occupational asthma. More than 80 different occupational asthma inducers have been identified. Occupational asthma develops when, on contact with the substance, the smooth muscles surrounding the bronchioles contract. This reduces the cross-sectional area of the bronchiole and restricts the flow of air—remember Poiseuille's Law.

Diffuse alveolar damage (DAD), clinically termed adult respiratory distress syndrome (ARDS), results when the cells lining the alveoli (pneumocytes) and alveolar capillaries (endothelial cells) allow protein-rich fluid to leak into the tiny spaces between the capillary and alveolus. This leads to the destruction of type I pneumocytes, inflammatory responses, and eventual pulmonary fibrosis (the formation of fibrous tissue in the lung), which may impair alveolar function.

Pulmonotoxicities resulting from the inhalation of mineral "dusts" are termed pneumoconioses. If, on inhalation, the mineral particulates enter the alveoli, they may stimulate the formation of pulmonary fibrosis. This decreases or eliminates the functional capacity of the affected alveoli.

In the United States, carcinoma of the lung is the most common cause of death from cancer. Carcinogens (cancer-causing chemicals) are causally linked to this form of pulmonotoxicity.

Examples

Among identified respiratory system irritants are ozone (O_3), nitrogen dioxide (NO_2), sulfur dioxide (SO_2), chlorine (Cl_2), and ammonia (NH_3). A decrease in the action of the mucociliary escalator is observed on exposure to cigarette smoke, ozone, and sulfuric acid. Clara cells, lining the terminal bronchioles, are highly vulnerable to ipomeanol, a fungitoxin produced by a mold that grows on sweet potatoes.

Occupational asthma has been causally linked to careers in agricultural harvesting, animal handling, food preparation, and woodworking. Pharmaceuticals (e.g., aspirin) and industrial chemicals (e.g., toluene di-isocyanate) are also implicated as pulmonotoxicants.

Pulmonary fibrosis, associated with DAD, is a pulmonotoxic response to the herbicide paraquat (1,1′-dimethyl-4,4′-dipyridilium). In addition to delayed toxic effects in the liver and kidneys, this weed killer produces interstitial alveolar fibrosis within 4–7 days following exposure. Paraquat can also be percutaneously absorbed, with subsequent distribution to the lungs. The higher concentrations of paraquat found in the lungs, as compared to other tissues, results from the active transport of paraquat across the alveolar cell membrane.

Depending on the causative agent, pulmonary fibrosis as linked to pneumoconioses may result from exposure to asbestos fibers (asbestosis), silicon dioxide particles (silicosis), coal dust that usually contains silica (anthracosilicosis), and beryllium (berylliosis). Although the pathogenesis of each pulmonotoxic disease differs slightly, the common pathology is fibrosis. Once inhaled into the alveolar sac, the "dust" is ingested by macrophages (large "scavenger" immune cells, monocytes). Altered by the presence of the "dust," the macrophages stimulate surrounding fibroblasts to secrete collagen, which leads to fibrosis.

There is a causal relationship between asbestos exposure and mesothelioma, a neoplasm involving the mesothelial cells that cover the lung (pleura). Not only is the incidence of mesothelioma high for asbestos workers, it is also found in the wives of asbestos workers, most likely a result of exposure when washing their husbands' contaminated clothes.

Carcinoma of the lung kills more than 100,000 persons in the United States each year. Approximately 90% of these lung cancers are a direct consequence of cigarette smoking. Lung neoplasms are also linked to exposure to arsenic, chromium, nickel, and coke oven emissions.

Evaluating Pulmonotoxicity

Respiration involves two separate but intimately related processes, each of which can be evaluated. The first of these processes includes the muscles of respiration, air passageways, compliance (elasticity) of lung tissues, and in general those structures and functions that facilitate the exchange of air between the lungs and the external environment. The total volume of air contained in the lungs is termed slow vital capacity (SVC). This volume can be measured with a spirometer by performing an SVC test. For the SVC test, subjects completely fill their lungs with a deep breath, then blow into a small, handheld device (spirometer) that measures the volume of air. A second test, the forced vital capacity (FVC) test, determines not only the SVC but also the volume of air expelled as a function of time. Decreases in the SVC or FVC usually indicate impaired pulmonary function (e.g., obstructions in air passageways).

Chest x rays, CAT scans, and MRIs are noninvasive approaches to detecting pulmonary pathologies.

Additional information is obtained by visually inspecting or sampling (e.g., biopsy) the larger passageways. Bronchoscopy is performed with a small fiber-optic instrument, a bronchoscope, inserted through the oral cavity and into the tracheobronchial region.

The second process ensures that the air (O_2) that enters the lungs as a result of "breathing" actually enters the blood for distribution to the tissues where it is used during cellular respiration. Arterial blood gases (ABGs) are good indicators of O_2 and CO_2 transport across the alveolar-endothelial membranes. When considering pulmonotoxicity, remember that normal "breathing" may not always result in sufficient "oxygenation" of tissues.

Table 7-1. Additional examples of target organ toxicity

Organ	Toxicant	Mechanism	Toxicity
Heart	Fluorocarbons (Freon)	?Sensitizes heart to epinephrine	Decreased heart rate, contractility, and conduction
	Carbon monoxide (CO)	Interferes with energy metabolism	Myocardial infarction, increase or decrease in heart rate
	Cobalt (Co)	Competes with Ca^{2+}	Heart failure
Testis	Lead (Pb)	?Mutations in sperm	Decreased male fertility, increased spontaneous abortions in wives
	Carbon disulfide (CS_2)	CNS effect on ejaculation	Reduced sperm counts
Ovary/uterus	Solder fumes	?Unknown	Increased spontaneous abortions
	Polycyclic aromatic hydrocarbons (PAH)	?Unknown	Damaged oocytes
Eye	Busulfan (chemo-therapeutic agent)	?Alters mitosis in lens cells	Formation of cataracts
	Methanol (CH_3OH)	Produces optic atrophy	Permanent visual impairment, blindness

Other Examples of Target Organ Toxicity

Cellular specialization in multicellular organisms has led to an array of different cell types. Virtually every organ in the human body, because of the unique assemblages of cells and tissues, is capable of exhibiting a different type of target organ toxicity. Table 7-1 contains additional examples of target organ toxicity.

Review Questions

1.   Which is *not* a true statement about target organ toxicity?

A.   It is defined as the adverse effects as manifested in specific organs of the body.
B.   Toxicity is unique for each organ.
C.   Toxicity may be enhanced by distribution features that deliver a high concentration of the toxicant to the organ.
D.   A single toxicant may have several target organs.
E.   The highest concentration of the toxicant is always found in the target organ.

2.   Examples of hematotoxicity include:

A.   Thrombocytopenia
B.   Microcytic hypochromic anemia
C.   Proteinuria
D.   A and B
E.   A, B, and C

3.   Hepatotoxicity is routinely evaluated by:

A.   Noninvasive testing of urine for particular enzymes associated with normal liver function
B.   PAH clearance test
C.   Testing clot formation times
D.   A and B
E.   A and C

4.   The mechanisms of neurotoxicity may include:

A.   An interruption in synaptic transmission
B.   The binding of neurotoxins to postsynaptic receptors
C.   The presence of nephritic syndrome
D.   A and B
E.   A, B, and C

5.   Glial cells function to:

A.   Transport nutrients to CNS neurons
B.   Produce myelin
C.   Provide a structural framework in the CNS
D.   A and B
E.   A, B, and C

6.   All of the following are noninvasive diagnostic tests used to assess neurological function except:

A.   Cranial nerve tests
B.   Sensory function tests
C.   Motor neuron function tests
D.   Patient history
E.   Cerebrospinal fluid tests

7.   Some individuals show little or no response on exposure to a chemical; however, on exposure at a later time they exhibit this delayed hypersensitivity reaction.

A.   Allergic contact dermatitis
B.   Chloracne

C. Chronic dermatitis
D. Hyperpigmentation
E. Irritant contact dermatitis

8. Pulmonotoxicity that results when the cells lining the alveoli allow protein-rich fluid to leak into the tiny spaces between the capillary and the alveolus:

A. Anthracosilicosis
B. Carcinoma of the lung
C. Diffuse alveolar damage
D. Occupational asthma
E. Pneumoconiosis

9. Discuss target organ toxicity as found in the eye, heart, and reproductive system.

10. Define the following abbreviations: ABG, ARDS, ATN, BUN, CAT, CBC/DIFF, CSF, DAD, EEG, EMG, GFR, HCT, Hb, MRI, PAH, PUVA, FVC, and SVC.

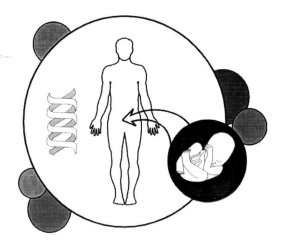

Chapter 8

Teratogenesis, Mutagenesis, and Carcinogenesis

Objectives

- Define teratogenesis, mutagenesis, and carcinogenesis

- Describe the relevance of replication, transcription, and translation to teratogenesis, mutagenesis, and carcinogenesis

- Summarize the mechanism of action for teratogens, mutagens, and carcinogens

- Discuss examples of known teratogens, mutagens, and carcinogens

Keywords

agenesis
Ames assay
aneuploidy
anticodons
atresia
base analogues
base substitution
cancer
carcinogenesis
carcinogens
cellular division
centromere
chromosome
codon
deoxyribonucleic acid (DNA)
developmental syndromes
diploid
division failures
dysraphic anomalies
ectopia
embryogenesis
embryolethality
epigenetic
fetal alcohol syndrome (FAS)
frameshift
gene
genetic code
genotoxic
germ cells

Keywords *(continued)*

haploid

histogenesis

hypoplasia

initiation

karyotype

meiosis

metaphase

mitosis

monosomy

morphogenesis

mutagenesis

mutagens

nucleic acids

nucleotides

oogenesis

organogenesis

point mutation

polyploidy

procarcinogen

promotion

purines

pyrimidines

replication

ribonucleic acid (RNA)

somatic cells

spermatogenesis

teratogenesis

teratogens

teratology

transcription

translation

trisomy

Introduction to Teratogenesis, Mutagenesis, and Carcinogenesis

Toxicants, in addition to their effects on target organs, may react with or modify the molecules of life. Once incorporated into these molecules, deoxyribonucleic acid (DNA) or ribonucleic acid (RNA), toxicants can have dramatic effects on the exposed individuals or their unborn offspring.

The perpetuation of life depends on the reproduction of cells, or cellular division. This is true for somatic cells (i.e., body cells), which reproduce over the life span of an individual, and germ cells (i.e., gametes), such as ova and spermatozoa, which form and subsequently join together during fertilization to facilitate continuation of the species. An understanding of the pathways by which genetic information is perpetuated (replication) and expressed (transcription and translation) is essential to comprehending the mechanisms of teratogenesis, mutagenesis, and carcinogenesis.

Teratogenesis (G. *teras*, monster; *-genesis*, origin) is the origin or production of malformed fetuses. Teratogens alter normal cellular differentiation or growth processes, which results in a malformed fetus (birth defects). Mutagenesis (L. *mutare*, to change) is the production of a mutation or change in the genetic code. On cellular division, these changes are passed on to "daughter cells." Carcinogenesis (G. *karkinos*, cancer) is the formation of cancer, including carcinomas and other malignant neoplasms. The substances responsible for causing mutations and neoplasms are termed mutagens and carcinogens, respectively.

Unlike target organ toxicity that affects exposed individuals, teratogens (and many mutagens and carcinogens) leave a legacy of toxicity to future generations.

Replication

Mitosis

In humans, all somatic cells contain 23 pairs of chromosomes. Each parent contributes one set (haploid or "n") to the pair of chromosomes in a diploid somatic cell. The expression "2n" (diploid) refers to these 46 chromosomes (i.e., 23 pairs).

Mitosis, a type of cell division, results in the production of two daughter cells with exactly the same chromosome number (2n) as in the original parent cell. Mitosis ensures that the genetic content of each daughter cell will be identical to the parent cell. This results from replication, a process that duplicates the cell's DNA.

Once replicated, each set of DNA becomes tightly coiled, wound, and condensed to form two chromatids. These two "sister" chromatids join in a region called the centromere to form a bivalent or tetrad chromosome (i.e., colored bodies). The characteristic)(appearance of human somatic bivalent chromosomes actually represents the replicated DNA (i.e., sister chromatids) connected at the centromere, with each side of the chromosome,) and (, delineating a single set of the DNA (i.e., chromatids). During mitosis, bivalent chromosomes will separate at their centromere and, upon completion, each daughter cell will have "inherited" half of the)(.

Meiosis

The formation of male and female gametes, spermatogenesis and oogenesis, respectively, results from a second type of cellular division termed meiosis. Unlike mitosis, meiosis produces daughter cells that possess only one set of chromosomes. The resulting haploid gametes have only 23 chromosomes (n), half of the 23 pairs of chromosomes normally found in somatic cells.

Unlike mitosis, meiosis guarantees that the gametes (i.e., daughter cells) produced will *not* be identical in chromosome number or genetic content to the parent cell. The first of these differences ensures that at conception (i.e., sexual reproduction), when the haploid ovum (n) and the haploid spermatozoon (n) unite, the diploid number (2n) of chromosomes will be restored in the fertilized ovum. The second difference is the source of the genetic variation so evident in all organisms using sexual reproduction to perpetuate their species.

Karyotypes

Chromosomes are best viewed and photographed during metaphase, a stage of mitosis (prophase → *metaphase* → anaphase → telophase). Once photographed, chromosomes can be arranged as pairs in descending order of size (i.e., overall height,)(→ x) and centromere placement. Centromeres may be found in the middle, as in metacentric chromosomes (e.g.,)(), or toward the ends of sister chromatids, as in acrocentric chromosomes. Once arranged, the systematized array of chromosomes is called a karyotype or karyogram (G. *karyon*, nucleus). Karyotypes are an invaluable aid in diagnosing chromosomal genetic anomalies.

Transcription and Translation

DNA occurs as a double strand of nucleotides (i.e., nucleic acid, sugar, and phosphate) in the form of a spiral ladder or double helix. There are two families of nucleotides, the pyrimidines (cytosine, thymine, and uracil), composed of a single carbon-nitrogen ring, and purines (adenine and guanine), composed of two carbon–nitrogen rings. Each rung or step of the ladder results when two of four possible nucleic acids (adenine, cytosine, guanine, and thymine) form a bond between the nucleotide sequences on each strand. Complementary base pairing always involves bonding between the nucleic acid bases adenine and thymine (A-T), and guanine and cytosine (G-C). The many different sequences of nucleic acid bases in each strand constitute the genetic code.

A gene represents a region or sequence of bases that contains the "blueprint" for a protein. (Remember that proteins are composed of amino acids.) A gene must code for a specific sequence of amino acids. Human DNA is composed of about 50,000–100,000 genes, which contain a total of about 6 billion base pairs.

With four different bases, and a code for a single amino acid consisting of a sequence of three bases called a codon, a total of 64 (4^3) codons are possible. This is more than enough codes to represent the 20 amino acids available to "build" proteins (Figure 8-1). Codons also serve as start and stop signals in the process of making proteins.

FIRST BASE ⇓	SECOND BASE								THIRD BASE ⇓
	U		C		A		G		
U	UUU	Phe	UCU	Ser	UAU	Tyr	UGU	Cys	U
	UUC	Phe	UCC	Ser	UAC	Tyr	UGC	Cys	C
	UUA	Leu	UCA	Ser	UAA	STOP	UGA	STOP	A
	UUG	Leu	UCG	Ser	UAG	STOP	UGG	Trp	G
C	CUU	Leu	CCU	Pro	CAU	His	CGU	Arg	U
	CUC	Leu	CCC	Pro	CAC	His	CGC	Arg	C
	CUA	Leu	CCA	Pro	CAA	Gln	CGA	Arg	A
	CUG	Leu	CCG	Pro	CAG	Gln	CGG	Arg	G
A	AUU	Ile	ACU	Thr	AAU	Asn	AGU	Ser	U
	AUC	Ile	ACC	Thr	AAC	Asn	AGC	Ser	C
	AUA	Ile	ACA	Thr	AAA	Lys	AGA	Arg	A
	AUG	Met or START	ACG	Thr	AAG	Lys	AGG	Arg	G
G	GUU	Val	GCU	Ala	GAU	Asp	GGU	Gly	U
	GUC	Val	GCC	Ala	GAC	Asp	GGC	Gly	C
	GUA	Val	GCA	Ala	GAA	Glu	GGA	Gly	A
	GUG	Val	GCG	Ala	GAG	Glu	GGG	Gly	G

Figure 8-1. The genetic code. A sequence of three nucleotide bases in the messenger RNA (mRNA) codon for the code for a specific amino acid. Note that there is a total of 64 (4^3) possible mRNA codons and that most amino acids are coded for by more than one codon.

DNA never leaves the nucleus, and a copy of the DNA "blueprint" must be made for export to ribosomes in the cytoplasm where actual protein synthesis occurs. An additional nucleic acid, RNA, functions as the messenger molecule between the nucleus and the cytoplasm. During a process called transcription, the double strands of DNA unzip along the bonds between complementary base pairs—the rungs on the ladder (Figure 8-2). This exposes the nucleotide sequence on each strand.

Specific enzymes facilitate the construction of a "copy" of each DNA strand, which means that new complementary base pairs are formed in the unzipped region. The RNA copy of DNA is almost identical to the original unzipped nucleotide sequence except for two differences. First, in RNA the nucleic acid uracil replaces thymine, so base pairing occurs between RNA's uracil (not thymine) and DNA's adenine, and RNA's adenine and DNA's thymine. Second, the five-carbon sugar deoxyri-

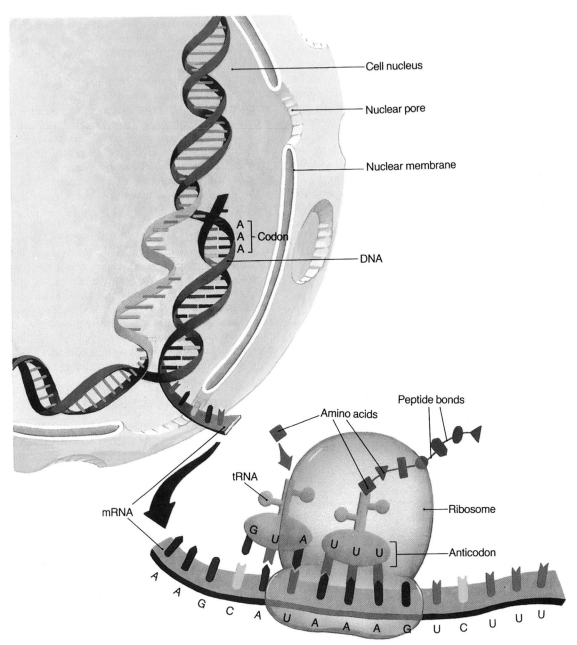

Figure 8-2. Overview of transcription (DNA → mRNA) and translation (mRNA → protein). (From V. C. Scanlon and T. Sanders, *Essentials of Anatomy and Physiology*, 2nd edition. F. A. Davis, 1995. Reprinted by permission.)

bose present in DNA (the D in DNA) is just ribose in RNA (the R in RNA). The newly formed RNA nucleotide sequence separates from the DNA template to form a single strand of RNA designated as messenger RNA or mRNA (Figure 8-2).

Transcription completed, the DNA template is transcribed to mRNA, and the two strands of DNA that were previously unzipped zip back together to once again form the double helix. The newly formed mRNA leaves the nucleus through nuclear pores to enter the cytoplasm, where it becomes the blueprint for protein synthesis.

Messenger RNA (mRNA) and two additional forms of RNA, transfer (tRNA) and ribosomal (rRNA or ribosomes), already present in the cytoplasm, are all involved in translation, the next phase of protein synthesis. Each form of RNA plays a role in translation: mRNA is the blueprint, rRNA is the factory (ribosome), and tRNA brings specific amino acids to the ribosomal protein factory.

Translation begins when mRNA arrives at the ribosome or rRNA (Figure 8-2). Since amino acids by themselves cannot interpret the sequence of nucleic acid codons present in mRNA, they must attach to anticodons present in tRNA. Transfer RNA performs two functions: (1) it attaches to a single specific amino acid that corresponds to the anticodon, and (2) it recognizes the correct mRNA codon. During translation, tRNA anticodons temporarily bond to the mRNA codons. One by one, in a sequence specified by mRNA, amino acids are linked together by the ribosome to form a protein. Once formed, the protein leaves the ribosome to be used by the cell or transported outside the cell.

Teratogenesis

Teratology is the study of developmental anomalies in fetuses. Worldwide, congenital anomalies are present in 2% of all newborns. Although animal studies indicate numerous chemical, physical, and biological teratogens, there are relatively few proven teratogens in humans. For ethical reasons, direct toxicity testing involving suspected teratogens in humans is not an option. Instead, researchers must rely on population surveys, retrospective studies, and investigations of reported adverse effects of suspected teratogens. Evidence from these approaches enables descriptive toxicologists and epidemiologists to infer "associative" relationships.

Through elaborate and complex pathways, the fertilized ovum (i.e., zygote) develops to form three germ layers—the ectoderm, mesoderm, and endoderm. On further differentiation, these germ layers give rise to all the specialized parts of the body (Figure 8-3). Teratogenesis results when these pathways are blocked, slowed, or in other ways altered. The resulting anomalies involve *in utero* malformations in the development of a body region, organ, or part of an organ.

The timing of exposure to teratogens is important because of the sequenced development of structures (e.g., CNS, heart) during early embryogenesis (formation of the embryo). There is a narrow time window during which exposure to a teratogen coincides with histogenesis (the formation of tissues) and organogenesis (the formation of organs). Exposure during this time window may result in: (1) malformation of the organ, (2) retardation or delayed formation of the organ, or (3) death of the embryo or fetus.

The most critical period for teratogenic activity is during the first 8 weeks of gestation, referred to as the embryonic stage. During this time there is maximal sensitivity to the development of morphological abnormalities in

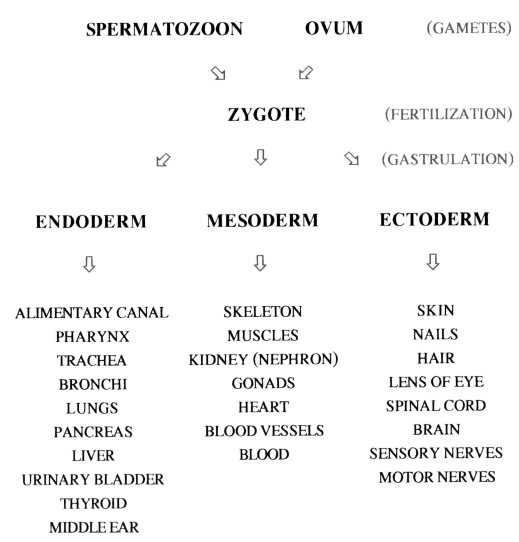

Figure 8-3. The origin of specialized tissues and organs in the human body.

response to teratogenic agents (Figure 8-4). Exposure to teratogens during the fetal stage (weeks 9 through 38) rarely results in major errors of morphogenesis (G. *morphe*, form or shape). However, these later teratogenic insults may result in neoplasms (i.e., cancer), organ or organ system dysfunction, and behavioral and developmental anomalies. Delayed functional maturation is evident in the CNS, which does not mature until several years after birth.

Typically, the outcomes related to errors in morphogenesis are dose dependent. At high doses, teratogens cause serious errors in morphogenesis. The severity of these errors usually results in embryolethality, which is death of the fertilized ovum or the embryo during the first 8 weeks (embryonic stage). An estimated 15–25% of all conceptions result in spontaneous abortion during the first 8 weeks. These percentages probably underestimate teratogenic embryolethal-

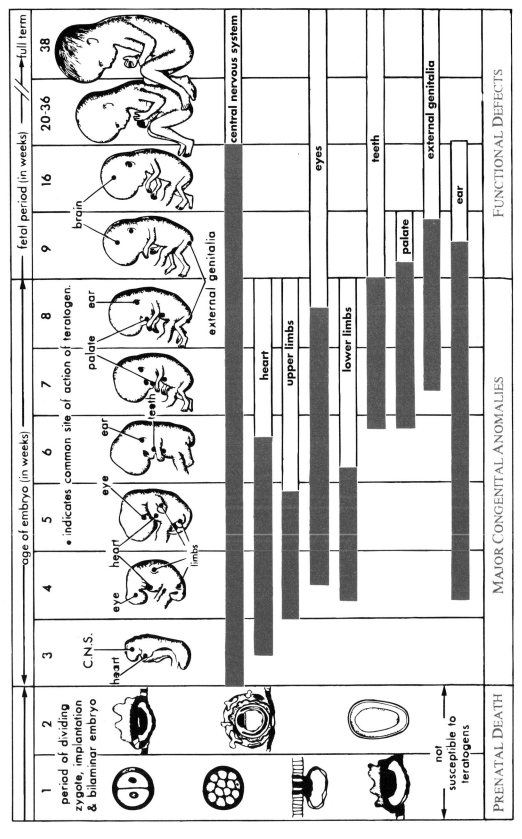

Figure 8-4. Sensitivity of tissues, organs, and organ systems to teratogens during embryogenesis. (From K. L. Moore and T. V. N. Persaud, *The Developing Human: Clinically Oriented Embryology*, 5th edition. W. B. Saunders, 1993. Reprinted by permission.)

ity, as embryonic death often goes unnoticed or presents as heavy menstrual bleeding.

Only over a narrow range of doses—not so high as to cause embryolethality and not so low as to have no effect—teratogens have deleterious effects on histogenesis or organogenesis. Examples include: agenesis, the complete absence of an organ (e.g., unilateral kidney agenesis); hypoplasia, reduced size of all or part of an organ (e.g., microcephaly); dysraphic anomalies, failure of apposed structures to fuse (e.g., spina bifida); division failures (e.g., syndactyly, the fusion of fingers); atresia, incomplete formation of a lumen (e.g., esophageal atresia); ectopia, presence of an organ outside its normal site (e.g., ectopic heart, with the heart outside the thoracic cavity); and developmental syndromes that involve multiple, but related, anomalies (e.g., fetal alcohol syndrome).

Mechanisms of teratogenic action must involve some form of cytotoxicity. Given the high rates of cellular division during histogenesis and organogenesis, any teratogen that interferes with the processes of cell division, replication, transcription, and translation may cause an error of morphogenesis. Descriptive toxicologists have identified numerous teratogens, but the biochemical mechanisms by which these agents have a teratogenic impact on humans, as well as other species, for the most part remain unknown. The following three examples, two drawn from pharmaceuticals and one from dietary preference, illustrate the specificity of teratogenesis. They also reveal the uncertainty of our understanding of the mechanisms of teratogenic action.

Thalidomide

In 1956, thalidomide was introduced in Europe as a sedative drug to alleviate the symptoms of "morning sickness," such as nausea and vomiting during the first trimester of pregnancy. The Food and Drug Administration (FDA) refused to approve the use of thalidomide in the United States until additional studies on its safety were completed. Meanwhile, evidence was accumulating about the safety of the drug. Of particular concern was the sudden increase in limb deformities in Germany and England, which were subsequently linked to the maternal use of thalidomide during the first trimester. The drug is now known to be a potent teratogen and the causative agent for numerous errors in morphogenesis involving limbs, ears, and the heart.

Phocomelia (G. *phoko*, seal; G. *melos*, limb) and amelia (i.e., without a limb or limbs), small or missing ears, and heart anomalies were among the teratogenic effects observed in over 7,000 children born to women who took thalidomide for therapeutic purposes early in pregnancy. Exposures to even as little as one tablet during the crucial time windows when limbs, ears, and the heart are forming are known to result in teratogenesis (Figure 8-5).

It is interesting that thalidomide does not exhibit teratogenic properties (i.e., phocomelia or amelia) in toxicity tests in which mice and rats are the experimental animals. However, in primate species (such as marmosets) teratogenesis is observed (Figure 8-6). The implications of generalizing conclusions from nonhuman toxicity testing are self-evident. After three decades of extensive research, the pre-

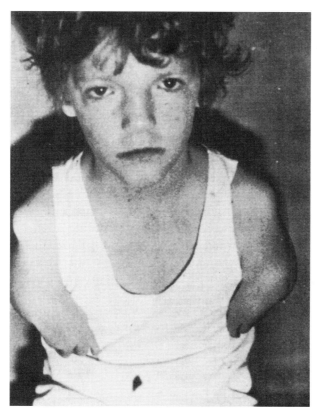

Figure 8-5. Phocomelia in child exposed *in utero* to thalidomide, after her mother took the drug during critical stages of limb bud development. (From E. Rubin and J. L. Farber, Eds., *Pathology*, 2nd edition. J. B. Lippincott, 1994. Reprinted by permission.)

cise mechanism of toxicity is still unknown. Although thalidomide was withdrawn from the market in 1961, its pharmaceutical value in the treatment of HIV/AIDS is currently being investigated.

Diethylstilbestrol

Between 1940 and 1971 over 2 million women were given diethylstilbestrol (DES), a potent synthetic estrogen (antiabortifacient) prescribed as a pharmaceutical treatment for high-risk pregnancies. DES is associated with a number of genital-tract anomalies present at birth and has also been implicated as a "delayed" teratogen.

Unlike thalidomide, whose teratogenic effects are manifest at birth, the delayed teratogenic actions of DES are observed in daughters, most commonly between the ages of 17 and 22, whose mothers took DES during pregnancy. The delayed teratogenesis includes an uncommon neoplasm (i.e., new growth, cancer), called a *clear-cell adenocarcinoma*, found almost exclusively in women exposed *in utero* to DES.

Although animal studies involving mice and rats show a dose-response relationship, the mechanism of teratogenesis is unknown. That teratogenic effects may be observed two decades after birth is sobering. The importance

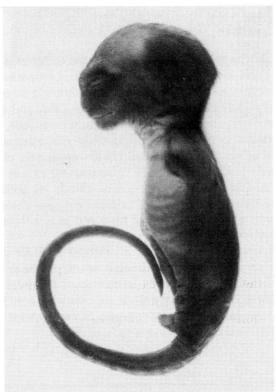

Figure 8-6. Thalidomide effects in fetal marmosets exposed to the drug between days 38 and 52 of gestation. Left, unexposed; right, exposed. (From W. G. McBride and P. H. Vardy, Pathogenesis of Thalidomide Teratogenesis in the Marmoset (*Callithrix jacchus*): Evidence Suggesting a Possible Trophic Influence of Cholinergic Nerves in Limb Morphogenesis, *Dev. Growth Differ.*, 25:361–373. Blackwell Science, 1983. Reprinted by permission.)

of designing long-term teratogenic studies again is self-evident.

Fetal Alcohol Syndrome

The consumption of alcoholic beverages during pregnancy results in a number of morphological and developmental anomalies. Among the observable effects are a thin upper lip, microcephaly (i.e., small head), micrognathia (i.e., small jaw), an underdeveloped philtrum (i.e., depression on upper lip), intrauterine growth retardation (IUGR), and CNS dysfunction. These anomalies are termed fetal alcohol syndrome (FAS).

Although the prevalence (existing cases) of FAS in Europe and the United States is 1–3 per 1,000 live births, in some populations with high rates of alcoholism the incidence (new cases) is 20–150 per 1,000. The greatest danger results from heavy alcohol consumption during the first trimester. Approximately 30–50% of women consuming more than 450 mL (1 pint) of "whiskey" per day will give birth to a child with FAS. In spite of extensive research, the teratogenic mechanism remains unknown.

Mutagenesis

Mutagens may affect either germ cells or somatic cells. Depending on the cell types affected, mutagenesis can have two very different outcomes. Somatic cells, unlike germ cells, are not the cells that contribute genetic information to the next generation. Mutations that occur in somatic cells only produce mutagenic, and often carcinogenic, effects in exposed individuals. Future generations are affected only when mutagens cause genetic damage to germ cells. Mutagens are capable of producing different types of genetic damage, ranging from small-scale point mutations in a single gene to large-scale changes involving whole chromosomes.

Point Mutations

The "simplest" type of genetic damage results when there is a mutation in the DNA sequence. Referred to as a point mutation, this mutation represents a change in the chromosome involving a single nucleotide (base) within the gene. These changes may result in the substitution, deletion, or insertion of a base.

A base substitution occurs when a nucleotide is substituted for a normally occurring base. If the substituted base does not alter the amino acid coded for in that position, then it will have *no effect* on the amino acid (Figure 8-7). This outcome is possible since each amino acid is coded for by more than one codon (see Figure 8-1). Two additional outcomes, *missense* and *nonsense*, result when the mutated triplet codon codes for a different amino acid or signals stop, respectively (Figure 8-7).

Although stated to be a "simple" type of genetic damage, point mutations result in complex mutagenic effects. Missense, resulting from base substitution, is responsible for sickle-cell disease. Substitution of the nucleotide adenine for thymine changes the codon for the sixth amino acid from CTC (glutamic acid) to CAC (valine). This "simple" substitution alters the amino acid sequence present in two of the four peptides that together form hemoglobin, producing RBCs that are structurally and functionally abnormal.

The deletion or insertion of a nucleotide results in an effect called a frameshift. On the deletion of a single nucleotide, the remaining bases must shift one position to fill the space once occupied by the now deleted base. The insertion of a nucleotide means that all bases must shift to allow for the added base. The effects of base deletion and insertion on protein translation are complex, since the translation reading frame has been shifted. Frameshifts may cause extensive missense or immediate nonsense (Figure 8-7). Frameshifts also result when entire triplet codons are deleted from or inserted into a nucleotide sequence.

Several physical and chemical mutagens are responsible for point mutations (Table 8-1). Base analogues are molecules whose chemical structure closely resembles pyrimidines and purines. These molecules (e.g., 5-bromouracil) may be substituted for normal bases during DNA synthesis (i.e., replication). The substituted base will be transcribed to mRNA during transcription. The base analogue substitution will have its ultimate impact during translation (i.e., no effect, missense, or nonsense).

MUTATION	CODON / PROTEIN

NORMAL

UUU AAG UAU GGC UAA
Phe - Lys - Tyr - Gly - Stop

BASE SUBSTITUTION

No Effect

UUU AAG UAC GGC UAA
Phe - Lys - Tyr - Gly - Stop

Missense

UUU AAU UAU GGC UAA
Phe - Asn - Tyr - Gly - Stop

Nonsense

UUU AAG UAA GGC UAA
Phe - Lys - Stop

BASE DELETION OR INSERTION

A
⇧
UUU AGU AUG GCU AA·
Phe - Ser - Met - Ala •••

Frameshift with missense

⇩
UUU UAA GUA UGG CUA
Phe - Stop

Frameshift with nonsense

⇩
CODON INSERTION OR DELETION   UUU AAG GUU UAU GGC
Phe - Lys - Val - Tyr - Gly

AAG
⇧
UUU UAU GGC UAA
Phe - Tyr - Gly - Stop

Figure 8-7. Examples of base mutations. Comparison of the normal codon and coded amino acid with mutations in which base substitution, deletion, and insertion have occurred.

Table 8-1. The effects of selected physical and chemical mutagens

Mutagen	Effect on DNA/RNA	Type of mutation
Ultraviolet radiation	C, T, and U dimers that cause base substitutions, deletions, and insertions	No effect, missense, or nonsense
X rays	Breaks in DNA	Chromosomal rearrangements and deletions
Acridines (tricyclic present in dyes)	Adds or deletes a nucleotide	Missense or nonsense
Alkylating agents	Interferes with specificity of base pairing (e.g., C with T or A, instead of G)	No effect, missense, or nonsense
5-Bromouracil	Pairs with A and G, replacing AT with GC, or GC with AT	No effect, missense, or nonsense

Chromosome Aberrations

Large-scale mutations may affect both the structure and number of chromosomes. Chromosome structure is altered when chromatids break off or are unexpectedly exchanged, centromeres fail to form, or "ring" chromosomes form. When these mutations occur during gametogenesis, and the resulting gamete is involved in fertilization, the altered chromosome(s) can be transmitted to offspring.

Numerical chromosome abnormalities are characterized as aneuploid or polyploid. Aneuploidy (without a true set) results when there is one more, or one less, chromosome present. The terms trisomy and monosomy, as well as the corresponding numerical formulas $2n + 1$ and $2n - 1$, are used to indicate these aneuploid conditions. Recall that the diploid state contains 46 chromosomes; therefore, $2n - 1 = 46 - 1 = 45$, and $2n + 1 = 46 + 1 = 47$.

Polyploidy (many sets) occurs when an additional set or multiple sets of chromosomes are present. (In humans, some liver cells are normally tetraploid or 4n.) The additional chromosome sets result from the formation of diploid (2n), rather than haploid (n), gametes. For example, an abnormal diploid spermatozoon that fertilizes a normal haploid ovum results in a triploid ($3n = 69$) zygote, or if both the spermatozoon and ovum are diploid, a tetraploid ($4n = 92$) zygote results. As evidenced by many agriculturally important polyploid plants, polyploidy is an important source of genetic variation. However, polyploidy in humans always results in embryolethality or death of the fetus. In rare instances a neonate, in spite of defects present in nearly all organs, may survive for a few days.

As with structural abnormalities, multiple sets of chromosomes in gametes can potentially be transmitted to offspring. Most likely underestimated, chromosome aberrations contribute to the incidence of spontaneous abortion and numerous genetic disorders (Table 8-2).

Table 8-2. Mutagenic effects on human chromosomes

Type of abnormality	Genotype
Sex chromosomes (male)	
Jacob's syndrome	47, XYY
Klinefelter's syndrome	47, XXY
Sex chromosomes (female)	
Turner's syndrome	45, XO
Metafemales	47, XXX
Autosomal trisomies	
Patau's syndrome	47, +13
Edward's syndrome	47, +18
Down's syndrome	47, +21
Deletions	
Cri-du-chat (female)	46, XX, 5p-
Cri-du-chat (male)	46, XY, 5p-

Note. Normal sex chromosomes (23rd pair) in males, XY; and females, XX. Arms (chromatids) of chromosomes are designated "p" (short arm) and "q" (long arm).

Ames Assay

Toxicity testing to determine which agents are mutagens is costly and time-consuming. This is particularly true when the test organisms are mammals, which are expensive to maintain and require longer experimental periods due to the overall slower rates of cell division. To reduce the cost and time, mutagenicity studies use bacteria and mammalian cell cultures. The Ames assay, named after Dr. Bruce N. Ames, is the most commonly used test for mutagenicity. The test uses the bacterium *Salmonella typhimurium*, which is cost-effective and time-conserving since it has rapid cell division. This assay also is valuable in identifying carcinogens, as about 90% of all known carcinogens exhibit mutagenic behavior in the Ames assay.

The Ames assay begins with a spe-cial strain of the bacteria that has a mutation in the gene coding for histidine synthesis. Since histidine is required for cell division, these bacteria will not multiply unless a back-mutation occurs in the histidine synthesis gene that again will permit histidine synthesis. Bacteria are exposed to the suspected mutagenic agent in a histidine-free growth medium (agar). Also present in the medium are microsomes from a rat's liver, should the test chemical require biotransformation to a reactive metabolite (i.e., mutagen). Finally, the rate of bacterial multiplication (i.e., colony growth) is monitored. If the suspected agent is *not* a mutagen, the bacterial colony will *not* grow. If the agent is a mutagen, the resulting mutations will permit histidine synthesis, and colony growth will take place.

Carcinogenesis

All cells possess an inherent rate of cellular division, one in which the rate of normal cell death is matched by the formation of new cells. Cancer, simply stated, occurs when there is uncontrolled proliferation of cells. Worldwide, cancer accounts for 4,800,000 deaths per year, a distant third cause of death behind cardiovascular disease (12,000,000), and diarrheal disease (5,000,000). In the United States, cancer is responsible for one-fifth of the total mortality, exceeded only by deaths from cardiovascular disease and stroke.

Remember, the World Health Organization (WHO) estimates that 90–95% of all cancers are "environmentally related." A few well-known examples are the relationship between lung cancer and cigarette smoking, cancer of the scrotum and occupational exposure of chimney sweeps to soot, and bladder cancer and exposure to dyes (i.e., aromatic amines) in textile workers.

Epidemiological observations supporting these relationships are based on a number of criteria, including the strength of the association and the consistency of the association under different circumstances. Even in epidemiological studies, the suspected causative agent must precede the effect (i.e., chronological sequence of the dose-response relationship), and an increase in dose must be paralleled by an increase in response. The specific mechanisms of carcinogenesis are for the most part poorly understood. This is due to the complex mechanisms involved in the development of cancer.

From experimental and epidemiological studies, we know that cancer results from a multistep process (Figure 8-8). Most frequently the complex process begins with a procarcinogen that is nonreactive. Only after the procarcinogen undergoes biotransformation (e.g., bioactivation) does it become

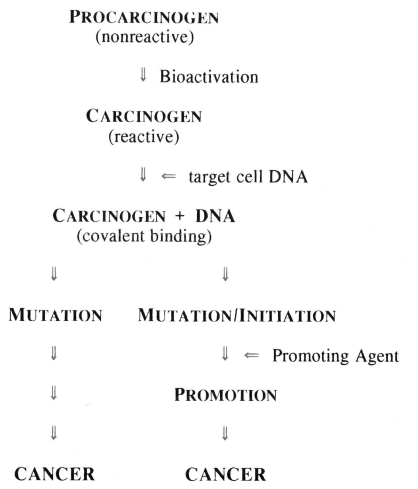

Figure 8-8. Proposed relationship among bioactivation, covalent binding, initiation, and promotion as related to carcinogenesis.

a carcinogen (i.e., reactive metabolite) that may covalently bind to DNA. The interaction between a carcinogen and DNA can produce the mutations leading to cancer formation.

According to the two-stage theory of carcinogenesis, in addition to bioactivation two other steps may be involved: initiation and promotion (Figure 8-8). Initiation involves a subtle alteration of DNA or proteins within target cells by the carcinogen. Promotion occurs when these altered cells, on exposure to a promoting agent, give rise to cancer. The promoting agent itself may be carcinogenic or noncarcinogenic.

Depending on their mode of action within the cell, carcinogens can be classified as genotoxic or epigenetic. Genotoxic carcinogens are *DNA reactive*; they act directly on DNA or the expression of DNA occurring during translation. Genotoxic mechanisms include errors in DNA replication, point mutations, and aberrations in chromosome structure and number.

Epigenetic carcinogens are *non-DNA reactive*. These carcinogens do not directly alter DNA; however, they are able to "affect" carcinogenic activity through numerous mechanisms. For example, epigenetic cocarcinogens may potentiate (increase) the activity of genotoxic carcinogens. Other epigenetic carcinogens act to increase cellular activity, modify the activity of hormones, or suppress the immune system, all of which may lead to cell proliferation (i.e., cancer).

Determining which agents are carcinogenic requires extensive testing. Initially, the Ames assay, as well as others, can be used to assess mutagenicity. Remember, about 90% of all known carcinogens exhibit mutagenic behavior in the Ames assay. If a compound is determined to be a mutagen, more costly and time-consuming carcinogenic studies using animal models are undertaken. Although there are hundreds of suspected human carcinogens, the list of actual or proven human carcinogens is quite short. It is estimated that in less than 5% of all cancers the cause is attributable to occupation, whereas about three-fourths of cancers result from diet (~50%) and tobacco (~25%). It is easy to see how rapidly we can approach the WHO estimate that over 90–95% of all cancers are "environmentally related"—diet, tobacco, and occupations are *all* controllable aspects of our environment.

Review Questions

1. All of the following statements about mitosis are true, except:

A. It produces daughter cells with exactly the same chromosome number as the parent cell.
B. All daughter cells are haploid.
C. Each daughter cell will be 2n.
D. It is used by somatic cells to reproduce.
E. It produces daughter cells that are identical in genetic content.

2. A karyotype:

A. Appears as a systematized array of chromosomes.
B. Is an invaluable aid in diagnosing chromosomal anomalies.
C. Results when chromosomes are viewed and photographed during telophase.
D. A and B
E. A, B, and C

3. Examine the following normal sequence of bases and resulting amino acid sequence:

AUG   AAG   UUU   GGC   GCA   UUG   UAA
Lys – Lys – Phe – Gly – Ala – Leu – Stop

Which answer best identifies the mutation in the base and amino acid sequence that follows?

AUG   AAG   UUU   GGU   GCA   UUG   UAA
Lys – Lys – Phe – Gly – Ala – Leu – Stop

A. Base substitution that has no effect.
B. Base substitution that causes missense.
C. Base substitution that results in immediate nonsense.
D. Base deletion that produces a frameshift and immediate nonsense.
E. Triplet insertion that results in missense.

4. Which is *not* a true statement about teratology?

A. Humans are often selected as the test organisms.
B. Teratogens may produce major errors of morphogenesis.
C. At low doses teratogens may produce embryolethality.
D. A and B
E. A, B, and C

5. Thalidomide, diethylstilbestrol, and ethanol are all examples of:

A. Carcinogens
B. Mutagens
C. Promoting agents
D. Teratogens
E. Translators

6. The greatest sensitivity of organs and organ systems to teratogens is:

A. Prior to fertilization.
B. During the preimplantation period.

C. During the first 8 weeks.
D. During the second and third trimester.
E. During the neonatal period.

7. Which term best describes a karyotype that contains 45 chromosomes?

A. Aneuploid
B. Monosomy
C. Tetraploid
D. Triploid
E. Trisomy

8. Base analogues are molecules whose chemical structure closely resembles pyrimidines and purines.

A. True
B. False

9. Diagram and discuss the relationship between bioactivation, covalent binding, initiation, and promotion as related to carcinogenesis.

10. Outline the sequence of events occurring during transcription and translation.

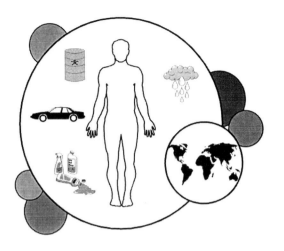

Chapter

9

Environmental Toxicants

bjectives

- Define environmental toxicants

- Recognize the contribution of environmental toxicants to worldwide morbidity and mortality

- Discuss representative categories of environmental toxicants, including examples

- Describe the mechanisms of toxicity within categories of environmental toxicants

eywords

Agent Orange
aliphatic alcohols
alkylbenzenes
anticoagulants
aromatic hydrocarbons
arsenic
benzene
beryllium
biological magnification
bipyridyls
cadmium
carbamates
carbon disulfide
carbon tetrachloride
cataractogenic
chelating agents
chlorinated aliphatics
chloroform
chlorophenoxy compounds
chromium
diabetogenics
dinitrophenol
dioxane
dioxin
dithiocarbamates
electromagnetic fields
environmental toxicants
ethyl alcohol

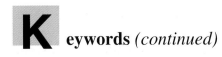

Keywords *(continued)*

ethylene glycol
fluoroacetic acid
fungicides
glycols
herbicides
hexachlorobenzene
insecticides
ionizing radiation
lead
Mee's lines
mercury
methyl alcohol
nickel
nickel itch
nonbiodegradable

nonphotodegradable
norbormide
organochlorines
organomercurials
organophosphates
pesticides
phthalimides
plastics
pyrinimil
radium
rodenticides
thermoplastics
thermosetting plastics
vasoconstrictors
warfarin

Introduction to Environmental Toxicants

Environmental toxicants are agents in our surroundings that are harmful to human health. Some substances, such as air and water pollutants, are recognized for their toxicity. No one would knowingly breath air polluted with sulfuric acid or drink water containing pesticides. However, other toxic agents equally harmful to human health (food additives and contaminants, bacteriotoxins, fungitoxins, phytotoxins, household products, and industrial chemicals) often go unrecognized for their serious toxicity.

Although man-made products (e.g., industrial chemicals) are thought to exhibit greater toxicity, historical data suggests that naturally occurring substances are a larger concern to human health. Without minimizing the danger of man-made products, consider the annual impact of bacteriotoxins alone on human morbidity and mortality. Secretory diarrhea, produced by toxins from bacteria (e.g., *Vibrio*, *Salmonella*, *Shigella*, and *Escherichia*), is responsible for the death of 5,000,000 persons worldwide each year—most of whom are children—who die from complications associated with dehydration and electrolyte imbalance.

This chapter synthesizes information on exposure, toxicokinetics, toxicodynamics, target organ toxicity, teratogenesis, mutagenesis, and carcinogenesis relating to specific categories of environmental toxicants. Examples are presented for each category, as well as pertinent information on their routes of absorption, modes of action, toxicokinetics, and clinical symptoms associated with toxicity. This is by no means an exhaustive listing; rather, it is intended to illustrate the principles of toxicology and explain by example the effects of environmentally hazardous substances on human health.

Pesticides

Pesticides are agents that destroy or repel unwanted organisms (i.e., pests; L. *pestis*, destruction, death, pestilence; *-cide*, kill). They are typically classified as to the organisms they destroy, as the terms fungicides (fungi), herbicides (plants), insecticides (insects), or rodenticides (rodents) imply. Animal pesticides are effective in eliminating unwanted animals because the pesticides use, to the extreme, the same toxicodynamics that produce toxicity in humans. Only differences in dose, exposure, and toxicokinetics—but not toxicodynamics—are usually evident. Neuronal transmission in cockroaches involves similar physiological phenomena as in humans.

Insecticides

Most insecticides are neurotoxicants that disrupt the transmission of a nerve impulse either as it passes along the axon or at the synapse. Insects exposed to neurotoxicants respond with twitching, weakness, and paralysis, which leads to death. Similar symptoms are also seen in humans.

Organophosphates. Parathion, diazinon, and malathion all inhibit cholinesterases (particularly acetylcholinesterase), the enzymes responsible for the degradation of the neurotransmitter acetylcholine. Failure to degrade acetylcholine released into neuronal synapses of the CNS and at myoneural junctions results in continued, repeated synaptic transmission that may lead to paralysis. In humans, absorption is through percutaneous, respiratory system, or digestive system routes. On distribution, organophosphates cross the blood-brain barrier to elicit CNS toxicity. Toxicants undergo phase I and phase II biotransformations in the liver and are subsequently eliminated. As neurotoxins, organophosphates involve most organs, including the gastrointestinal tract (nausea, vomiting), the respiratory system (excessive bronchial secretions), the cardiovascular system (decrease/increase in heart rate or blood pressure), skeletal muscles (weakness, paralysis), and the CNS (mental confusion, fatigue).

Carbamates. Like organophosphates, carbamates (aldicarb, carbaryl, propoxur) inhibit the enzymatic action of cholinesterases. The toxicant enters the body through the percutaneous, respiratory system, and digestive system routes. In humans, oral doses of as little as 3 mg/kg can result in toxicity, and fatalities due to carbamate toxicity have been reported. Biotransformation reactions rapidly break up (hydrolyze) the carbamate cholinesterase molecule, thereby reactivating cholinesterase. This explains the unusually short duration of carbamate-induced neurotoxicity. CNS and neuromuscular junction symptoms include nausea, vomiting, sweating, muscle weakness, and—in severe cases—convulsions.

Organochlorines. Dichlorodiphenyltrichloroethane (DDT) and other chlorinated organic insecticides act to stimulate or depress the CNS. The neurotoxicity of DDT is thought to result from membrane-altering processes that act to diminish the rate of repolarization, such as impaired transport of Na^+ and K^+ in the axon, and Ca^{2+} as it signals the release of neurotransmitters in the region of the synapse. Neurons that are not fully repolarized require less of a stimulus to initiate signal transmission; thus, affected neurons have increased sensitivity that leads to repetitive signaling. Primary routes of absorption differ for representative organochlorines. Case reports show there is less toxicity associated with cutaneous exposure, probably due to poor absorption through the skin. The more common route of absorption leading to toxicity involves the ingestion of DDT.

The toxicokinetic properties of organochlorines that make them good insecticides also are properties responsible for the banning of DDT. For example, as a result of high chemical stability, DDT and its breakdown products/metabolites persist in the environment. DDT's lipid solubility, coupled with a slow biotransformation rate, promotes accumulation in individual organisms, as well as in organisms farther up the food chain (i.e., biological magnification). Organochlorines are readily stored in fat and are slowly eliminated at a rate of about 1% per day. Complex biotransformation pathways include both phase I (dechlorination, demethylation) and phase II (glutathione conjugation) reactions, followed by elimination. Clinical symptoms associated with acute CNS toxicity include headaches, dizziness, tremors, and convulsions. Symptoms of chronic toxicity are loss of memory, personality changes, and a reduction in sperm count in males.

Herbicides

Herbicides act to eliminate unwanted plants (e.g., weeds) by interfering with hormonal systems that regulate growth or by promoting water loss (i.e., desiccation). Although effective in eliminating plants, most herbicides are weakly toxic to humans, most likely due to inherent differences in plant and animal cell structure (e.g., cell membrane) and function (e.g., biochemical pathways).

Bipyridyls. As herbicides, bipyridyls (e.g., diquat, paraquat) act to desiccate plants. Although these toxicants may be absorbed through percutaneous, respiratory, or digestive system routes, the capacity for paraquat to preferentially become concentrated in the lung tissues is well documented. On distribution to the lung, paraquat decreases gas exchange by damaging pneumocytes. This in turn decreases the transport of O_2 and CO_2 across alveolar cell membranes. Biotransformation of bipyridyls is poorly understood; however, elimination is via urinary and fecal routes. Clinical symptoms of paraquat toxicity include anoxia and coma, as well as damage to the lungs, liver, and kidneys. The ingestion of concentrated paraquat almost always leads to death.

Chlorophenoxy Compounds. 2,4-Dichlorophenoxyacetic acid (2,4-D) and 2,4,5-trichlorophenoxyacetic acid (2,4,5-T) are well-known chlorophenoxy compounds. Agent Orange, a 50:50 mixture of 2,4-D and 2,4,5-T, was extensively used during the Vietnam conflict. The herbicidal properties of these compounds promote uncontrolled plant growth that rapidly leads to plant death. Chlorophenoxy compounds are weakly toxic to humans, but a trace impurity (2,3,7,8-tetrachloro-dibenzo-p-dioxin or TCDD) that results from the manufacturing of 2,4,5-T is a powerful toxicant. Animal studies indicate that TCDD (or dioxin as it is commonly referred to in the popular media) is a potent toxicant recognized for its dermatotoxicity (chloracne) and teratogenic and carcinogenic properties. Chlorophenoxy compounds are absorbed through percutaneous or respiratory and digestive system routes. Little is known about their toxicodynamics or toxicokinetics, especially as related to biotransformation and routes of elimination. Symptoms associated with toxicity include sweating, scanty urine production (oliguria), peripheral neuropathies, muscle weakness, dizziness, headaches, vomiting, and fatigue. The LD_{50} for 2,4-D is about 300 mg/kg, with threshold TDs ranging from 50 to 60 mg/kg.

Dinitrophenol. In addition to its herbicidal use, 2,4-dinitrophenol (DNP) was once marketed as an over-the-counter antiobesity agent. Aside from pharmaceutical dosing, exposure may occur through percutaneous or respiratory and digestive system routes. Toxicity is due to DNP's capacity to inhibit ATP synthesis, which may lead to acute symptoms of tachypnea (rapid breathing), tachycardia (rapid heart rate), sweating and coma, and chronic symptoms of fatigue, anxiety, and weight loss. Of additional interest are the cataractogenic properties of DNP. Cataracts (loss of transparency of the lens of the eye) developed in more than 100 persons who used DNP as an antiobesity agent from 1935 to 1937. As is typical for many aromatic nitro and amino compounds, DNP is also a carcinogen. There is a paucity of information on toxicokinetic properties related to biotransformation and elimination of DNP.

Fungicides

Although many fungicides are no longer in use, a review of their toxic properties provides an important reminder of the general safety as related to the approximate 100 million pounds of fungicides used each year in the United States. Fungi, the intended victims of fungicides, are themselves responsible for producing some of the deadliest toxins (i.e., fungitoxins). For example, *Amanita phalloides*, the "death cap mushroom," contains potentially lethal phallotoxins, and aflatoxins (B_1) from *Aspergillus flavus* are well documented for their hepatocarcinogenic activity.

Hexachlorobenzene. Prior to 1960, hexachlorobenzene (HCB) was used to treat seed grain. This prevented fungal infestation in seeds before they were planted. In the late 1950s, about 4,000 Turkish citizens became seriously ill when they mistakenly ingested HCB-treated *seed* grain for *food* grain. Toxic responses include skin blisters, hepatomegaly (enlarged liver), and thyroidomegaly (enlarged thyroid), as well as arthritis, osteomyelitis (inflammation within bones), and osteoporosis (loss of bone density) in the hands. Oral ingestion is the obvious route of absorption. Little is known about the toxicokinetics of HCB; however, like organochlorines (e.g., DDT), HCB persists in the environment, is biomagnified, and has a comparatively long biological $T_{1/2}$ due to its slow rate of biotransformation.

Organomercurials. Mercurial compounds (e.g., methylmercury) were used to treat seed grains as recently as the 1970s. In two incidents, separated by thousands of miles, organomercurials were implicated in epidemic poison-ings. In Iraq the direct ingestion of treated grain was responsible for toxicity, whereas in New Mexico, treated grain was fed to hogs that were subsequently slaughtered, and persons consuming these hogs became ill. Again the digestive system was the route of absorption. Acute toxicity involves the gastrointestinal and renal systems; particularly evident is proximal tubule damage within the nephron.

Phthalimides. These potent fungicides (Captafol, Folpet) have chemical structures similar to thalidomide—a similarity that has caused controversy about their use (Figure 9-1). In spite of the oral LD_{50} values of 10,000 mg/kg observed in rats, uncertainty over mixed results from teratogenic and mutagenic studies in hamsters, mice, and rats has halted their use. The toxicodynamics of phthalimides are unclear; however, there is indication that they may interfere with enzymes. As dusts, emulsions, and sprays, these chemicals are primarily absorbed through respiratory system and percutaneous routes. Animal studies indicate phthalimides are rapidly biotransformed and eliminated in the urine and feces. Symptoms of acute toxicity include irritant and allergic contact dermatitis, and chronic toxicity, mutagenicity, carcinogenicity, and teratogenicity as evidenced from nonhuman animal studies.

Dithiocarbamates. Besides their use as fungicides (e.g., ethylene-bidithiocarbamate, or EBDC), some dithiocarbamate derivatives are used as insecticides (acetylcholinesterase inhibitors) and in the chemical and rubber industries. As evidenced in the chemical brand names Ferbam, Maneb, Nabam, and Zineb, dithiocarbamates contain metal ions, such as Fe^{3+}, Mn^{2+}, Na^+, and Zn^{2+},

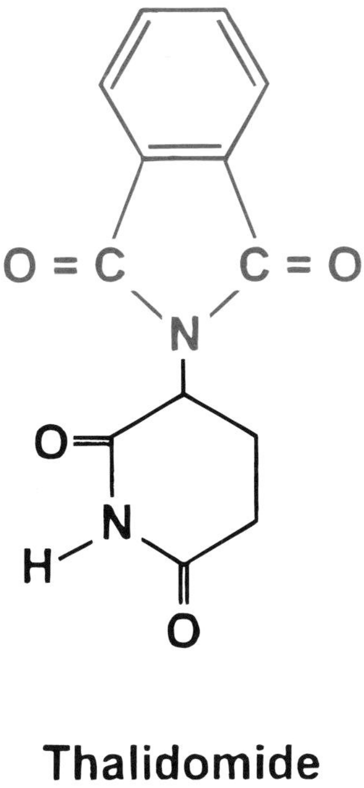

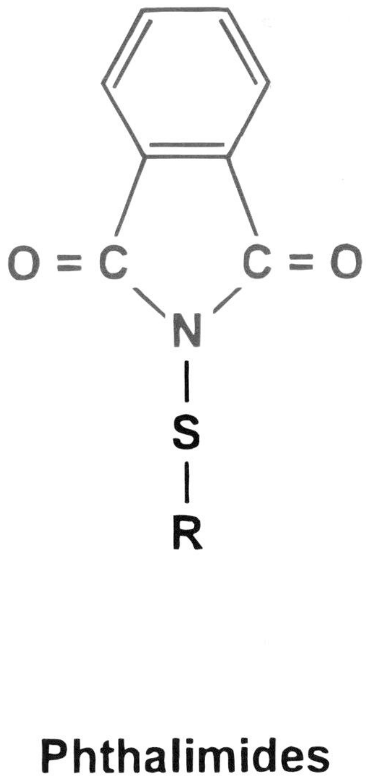

Thalidomide   Phthalimides

Figure 9-1. Similarity between the chemical structure of thalidomide (a teratogen) and phthalimide (a fungicide). "R" represents a variety of functional groups, each of which distinguishes a specific phthalimide fungicide.

respectively. Dithiocarbamates enter the body through percutaneous or respiratory and digestive system routes, and are rapidly distributed, biotransformed, and eliminated. In general, dithiocarbamates are of relatively low toxicity to humans. Depending on the associated metal ion, toxicity may include both irritant and allergic contact dermatitis and CNS depression. Animal studies give indication of dithiocarbamate's mutagenic and carcinogenic potential.

Rodenticides

Numerous vertebrate organisms (such as bats, coyotes, rabbits, skunks, and wolves) have at one time or another been considered pests. Rodenticides (agents lethal to rodents) are of particu- lar importance to human health, since rodents (e.g., rats and mice) often serve as vectors for the transmission of disease. Most recently, the emergence and spread of the hantavirus has been linked to a sudden increase in deer mice, which are carriers of the virus. Use of rodenticides in the deer mice population is just one of many solutions to control the spread of hantavirus. It is not surprising that, due to similar cellular processes in the two orders, Rodentia and Primates (both in the class Mammalia and phylum Vertebrata), there would be similar manifestations of toxicity, the severity of which is dose related. Examples of rodenticides include anticoagulants, inhibitors of cellular respiration, vaso- constrictors, and diabetogenics. Since most rodenticides are topically applied to palatable baits (e.g., seeds), the anticipated route of absorption is the gastrointestinal system. Aside from accidental percutaneous and respiratory system exposures in the manufacturing process, human toxicity often results from the ingestion of rodenticides, particularly the accidental ingestion by children.

Anticoagulants. Medically important anticoagulants (heparin, warfarin, aspirin) are routinely used in the treatment of thrombi (stationary blood clots) and emboli (transient blood clots). The ability of these drugs to prevent the aggregation of thrombocytes (platelets) effectively "thins" blood—an action that is exploited in their use as rodenticides. A common rodenticide, warfarin (coumadin), inhibits the synthesis of vitamin K. In the absence of vitamin K, liver cells are unable to produce prothrombin, a molecule vital to the completion of a series of cascading reactions that lead to clot formation. In humans, repeated exposure to warfarin is usually required to cause blood to thin, which may result in internal hemorrhaging, as seen in bruises, gastrointestinal bleeding, cerebrovascular accidents (strokes), and nosebleeds. Differences in anticoagulant toxicokinetics are apparent in plasma protein binding comparisons—warfarin (97%), aspirin (50–70%), heparin (trace %)—and biological $T_{1/2}$ comparisons: warfarin (40 hours), aspirin (3 hours), and heparin (1 hour).

Inhibitors of Cellular Respiration. Derivatives of fluoroacetic acid (sodium fluoroacetate, fluoroacetamide) are extremely toxic rodenticides. These chemicals block one of many enzymes involved in the Krebs cycle, impeding a major pathway responsible for ATP production. In the absence of the high-energy molecule ATP, cellular functions come to a halt. Clinical symptoms include nausea, vomiting, abdominal pain, increased heart rate, kidney failure, coma, and death (with threshold LDs of ≤10 mg/kg for sodium fluoroacetate).

Vasoconstrictors. The toxicodynamics of norbormide involve contraction of the smooth muscles surrounding peripheral blood vessels. Under normal conditions the contraction and relaxation of this musculature surrounding blood vessels is an important mechanism for regulating blood pressure. However, in rats, 5–15 mg/kg of norbormide elicit irreversible vasoconstriction, leading to ischemia (reduced blood flow to tissues) and necrosis (cell death), followed by death of the organism. These toxic responses are not seen in other vertebrate species, including even acute exposures of up to 300 mg in humans.

Diabetogenics. As exemplified by pyrinimil, diabetogenics interfere with glucose metabolism by exerting cytotoxic effects on β islet cells of the pancreas. The hormone insulin, produced by β cells, is necessary to facilitate the passage of glucose across the cell membranes of all body cells. The absence of glucose in the cell leaves the cell without the ready energy source needed to produce ATP via glycolytic pathways. On exposure to pyrinimil, there is an initial period of hyperglycemia (abnormally elevated blood glucose levels) due to the inability of glucose to move from circulating blood into the surrounding tissue cells. Hypoglycemia (abnormally decreased blood glucose levels) usually follows, with symptoms of light-headedness and urinary retention. Neurological disorders, both sensory and motor, may also be present.

Plastics

Polymers that can be shaped by pressure or heat to the form of a cavity or mold are termed plastics (G. *plastikos*, fit for molding). Two forms of plastics are generally recognized: thermoplastics and thermosetting plastics. Thermoplastics, which comprise over 80% of all plastics, can be remelted and remolded (e.g., polyethylene, polypropylene, polyvinylchloride, polystyrene), and as such are of interest to numerous recycling efforts. Plastics that once molded *cannot* be remelted and remolded are termed thermosetting plastics. Concern about plastics as environmental toxicants is twofold. First, many plastics—thermoplastics in particular—are nondegradable. They resist biological degradation (nonbiodegradable) and degradation from ultraviolet radiation (nonphotodegradable). The possibility is that plastics could persist in landfills or as roadside pollutants for hundreds of years. Although more than a fourth of all aluminum and paper are recycled in the United States each year, only about 1% of plastics are recycled. Second, attempts to incinerate some plastics (e.g., polyvinylchloride or PVC), to reduce their contribution to landfills, result in the production of toxic chemicals. One such toxic chemical, dioxane, may be absorbed via the respiratory system. In animal studies dioxanes are demonstrated carcinogens, most probably involving an epigenetic mechanism.

Metals

Although many of the 80 known metals are vital to normal physiological processes in humans (e.g., Fe, Mg, Zn); other metals, such as Pb, Hg, and Cd, are among the oldest toxicants known to humans. Metals are unique as toxicants—they are neither created nor destroyed by organisms, plants, or animals, because as chemical elements they cannot be degraded beyond their elemental states. In fact, in somewhat parallel roles (Mg in chlorophyll and Fe in hemoglobin), metals often accomplish their functional role by switching between valences (e.g., $Fe^{++} \leftrightarrow Fe^{+++}$) in their interaction with other molecules. They may be in the form of elemental metals (e.g., Hg), nonorganic salts (e.g., $HgCl_2$), or organic metal compounds (e.g., $Hg(CH_3)_2$).

Metals enter the body through digestive and respiratory routes. Urine is the most common route of elimination and nephrons are often the site of toxicity, as evidenced in the tubules. Metals may be directly excreted through the intestinal mucosa into the lumen of the digestive tract. The enterohepatic circulation of some organic metal compounds (e.g., methylmercury) serves to increase their $T_{1/2}$. Other mechanisms contribute to the elimination of metals, such as loss through breast milk, hair, nails, and exfoliating skin.

Many metals are considered essential to normal cellular activity. However, in excess they may cause toxic responses (Table 9-1). Metals are also used in medical diagnostics and treatments (Table 9-2). Pharmaceuticals are available to assist in the removal of toxic metals from the body. These chelating

Table 9-1. Essential metals

Metal	Function	Toxicity associated with excess
Co	Found in vitamin B_{12}	Polycythemia, cardiomyopathy
Cu	Synthesis of hemoglobin	Microcytic anemia
Fe	Erythropoiesis (formation of RBCs)	Liver and cardiovascular damage; if inhaled, silicosis-like symptoms
Mn	Enzyme potentiator	Manganese pneumonitis, CNS disorders
Mo	Enzyme cofactor	Anemia and diarrhea (animal studies)
Se	Enzyme cofactor	Neuropathies, dermatopathies, decreased fertility, teratogenesis

agents facilitate the elimination of toxic metals by forming a "metal–ion" complex that is more reactive and hence more readily eliminated (Table 9-3). Most chelating agents are nonspecific. Monitoring essential metal ions (e.g., Ca^{2+}) during the administration of chelating agents is necessary, since a reduction in blood levels may lead to pathologies (e.g., muscle and nervous system dysfunction).

Arsenic

The "toxic and tonic" value of arsenicals is controversial. Toxic properties of arsenic have been recognized for hundreds of years, and the medicinal use of arsenic tonics was common as recently as 100 years ago. Their use in insecticides, weed killers, and wood preservatives is of concern, since arsenic is toxic to a variety of organisms. On ingestion arsenic is

Table 9-2. Medically important metals

Metal	Medical use	Toxicity associated with excess
Al	Antacids, dialysis fluids	Dialysis dementia
Bi	Antacids, antisyphilitics	Nephropathies
Ga	Radiographic imaging	Dermatitis, gastrointestinal disorders
Au	Pharmaceuticals (rheumatoid arthritis)	Nephropathies
Li	Pharmaceuticals (depression)	Nephropathies, cardiopathies
Pt	Pharmaceuticals (chemotherapy)	Carcinogenesis, nephropathies

Table 9-3. Chelating agents

Agent	Metal toxicities treated
Edetate calcium disodium (CaNa$_2$-EDTA)	Cd, Cr, Co, Cu, Pb, Mn, Ni, Ra, Se, Te, U, V, Zn
British anti-Lewisite (BAL) (2,3-dimercaptopropanol)	Au, As, Bi, chromates, copper salts, Hg, Ni, Sb, W, Zn
Penicillamine (β,β^1-dimethylcystein)	Cd, chromates, Co, copper salts, Hg, Ni, Pb, Zn
Succimer (DMSA) (2,3-dimercaptosuccinic acid)	Pb (approved for children whose blood lead levels are greater than 45 μg/dl)

effectively absorbed by the gastrointestinal system and also may enter the body via the respiratory system. The kidneys are the primary pathway for the elimination of arsenic; however, it also is lost from the body as a result of desquamation (loss of stratified squamous epithelium—skin), in sweat, hair, and in the fingernails and toenails. Signs of chronic arsenic toxicity are evidenced in the nails by the presence of horizontal white bands (i.e., Mee's lines). Acute arsenic toxicity includes anorexia, hepatomegaly, possible cardiovascular failure, and death (threshold LD of >70mg). Chronic toxicity results in both CNS and PNS pathologies, including muscle weakness and loss of sensory perception.

Beryllium

This metal is released by the combustion of coal and in the manufacture of alloys, mainly those associated with aerospace industries. Beryllium is absorbed via the respiratory system, a route that, under chronic exposure, presents the lungs with sufficient concentrations of the metal to produce berylliosis, a disease in which the lungs decrease in size, become fibrotic, and may develop cysts (or honeycomb lung). The most common toxic effect associated with dermal exposure is allergic contact dermatitis. Epidemiological studies indicate that beryllium is a human carcinogen.

Cadmium

Used in manufacturing processes and in many household products, cadmium readily enters the body through the respiratory system. Once in the blood, it binds to large proteins (e.g., albumin) for distribution to tissues, primarily the kidneys. Cadmium has a long biological T$_{1/2}$, possibly 30 years. Toxicity associated with acute respiratory exposure may include pulmonary edema (accumulation of fluid in the lungs), whereas ingested cadmium may result in nausea, vomiting, and abdominal pain. Chronic exposures are linked to nephrotoxicity, most often affecting the tubules rather than the glomeruli. Recent epidemiological studies examining "Ni-Cad" (nickel–cadmium) battery workers in Britain and Sweden suggest a link between cadmium exposure and increased risk for developing cancer of the prostate and lung.

Chromium

The main sources of chromium exposure come from chromite ore mining and subsequent uses in the production of stainless steel, paint pigments, and wood preservatives, and in leather tanning. Of the many chromium valences (Cr^{2+} to Cr^{6+}), only the trivalent and hexavalent are thought to be of biological importance; however, toxicity is usually only associated with hexavalent chromium. The respiratory system is the major route of chromium absorption. A common renal symptom associated with the ingestion of chromium is acute tubular necrosis (ATN). Chromium is recognized as a contact allergen and carcinogen (cancer of the lung).

Lead

In 1815, Orfila recognized "poisoning by lead as the most important to be known of all those that have been treated of, up to the present time." Lead in ceramics, paints, and automobile exhausts is still recognized for its toxicity, particularly as it affects children. Absorbed through respiratory or digestive system routes, lead preferentially binds to RBCs for distribution to the tissues. Of interest is the storage of lead in bone, where its $T_{1/2}$ may exceed 20 years. Although slow, the elimination of lead occurs through the kidneys. At about 100 µg/dl (micrograms per deciliter of blood), toxicity symptoms include hematopathies, neuropathies, and nephropathies. Carcinogenesis—demonstrated in animals but not in humans—is also suspected. In children, lead encephalopathy (disease of the brain) may result in loss of appetite, ataxia, coma, and death.

Mercury

Much of the mercury in the environment originates from natural geological processes, such as the degassing of the earth's crust. As with other metals, toxicity may result from elemental mercury, nonorganic mercury salts, and organic mercury compounds. There are differences in the absorption of the various forms of mercury, with higher gastrointestinal absorption for organic mercury. Also, different effects on target organs for each form of mercury are evident, with nonorganic salts concentrating in the kidneys and organic mercury showing a preference for the brain. Elimination, again form dependent, is through urine or feces. Animal studies show that all forms of mercury cross the placental barrier, and it is most likely that since organic mercury crosses the blood-brain barrier in humans, it also moves across the human placenta. The tragic consequences of mercury toxicity were well illustrated in Minamata Bay, Japan, during the 1950s, where mercury released from a chemical factory entered the bay and contaminated the food supply. Toxicities resulting from exposure to methylmercury include neuropathies, nephropathies, teratogenesis, and mutagenesis.

Nickel

Used in alloys, batteries, coins, electronics, and food processing, nickel is absorbed primarily through the respiratory system, with less efficient absorption occurring in the digestive system. It readily binds to plasma proteins and is rapidly eliminated by the kidneys. In sensitized individuals, dermal exposures that occur when handling coins or wearing costume jewelry may result in allergic contact dermatitis—commonly called "nickel itch." Epidemiological studies have shown a strong association between occupational exposure to nickel and increased risk of developing lung cancer and cancer of the nasal cavities.

Organic Solvents

The toxicological parameters for all known carbon-based solvents cannot be fully described—there are just too many. However, representative categories (Figure 9-2), with relevant examples, illustrate the toxic effects of organic sol-

vents on human health. In general, two toxic responses are observed to result from exposure to organic solvents: (1) depression of the CNS and (2) irritation of tissues and membranes. The latter is expected, as lipids in the cell membrane are vulnerable to the solvent characteristics of organic molecules—organic solvents "defat" the membrane!

The ability of organic solvents to depress the CNS is exploited with intravenous and inhalational anesthetics. Since the blood-brain barrier (as well as the brain itself) is largely composed of lipids, effective anesthetics must have high lipid solubilities. In fact, the higher

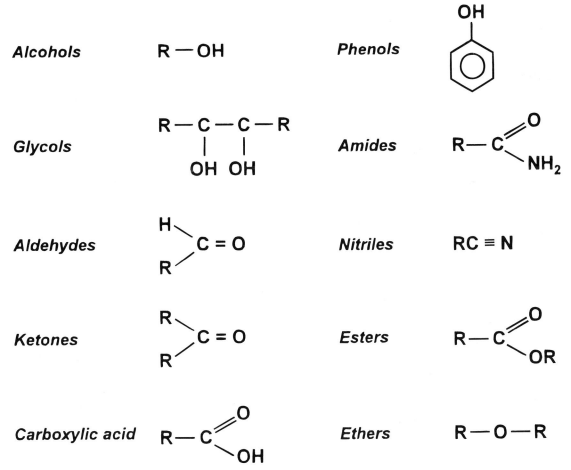

Figure 9-2. Representative organic chemicals.

the lipid solubility of an anesthetic, the higher its potency. Once in the CNS, organic solvents—like anesthetics—exert their depressant actions by interfering with neuronal signaling.

Although toxicity is often associated with occupational exposure to organic solvents, their ingestion by children is a serious concern—one-fourth of all pediatric poisonings involve organic solvents, usually from household products that contain hydrocarbons. Solvents are also potent nephrotoxins, hepatotoxins, and cardiotoxins.

Aliphatic Alcohols

Aside from water, ethyl alcohol (ethanol) is the most common solvent. Present in alcoholic beverages, ethyl alcohol enters the body by ingestion. Other exposures may result from contact with "gasohol," or its use as a chemical feedstock or in manufacturing. Ethyl alcohol crosses the blood-brain barrier and depresses the CNS. Ethyl alcohol's role as an occupational toxicant pales next to its role as an abused "drink"—each year thousands of innocent victims are killed by drivers under the influence of alcohol's CNS depressant effects. It is also a well-known teratogen (FAS) and carcinogen linked to oral, pharyngeal, laryngeal, esophageal, and hepatic cancers. Another aliphatic alcohol, methyl alcohol (methanol or wood alcohol), is easily absorbed through percutaneous or respiratory and digestive system routes. Although less inebriating than ethyl alcohol, methyl alcohol targets the neural cells in the retina of the eye to produce intra-axonal swelling. This may lead to visual system pathologies, including permanent blindness.

Chlorinated Aliphatics

Used as an anesthetic (controlled neurotoxicity) up to the early 1900s, chloroform ($CHCl_3$) is now recognized as a nephrotoxin, hepatotoxin, and cardiotoxin. Absorbed through the respiratory system, chloroform is readily metabolized to reactive metabolites that are toxic to the kidney, liver, and heart. Carbon tetrachloride (CCl_4), used as a solvent in dry cleaning, is highly hepatotoxic. One suggestion is that the toxic effects in the liver result from CCl_4 reactive metabolites that bind to and inactivate cytochrome P-450. Once inactivated, cytochrome P-450 is no longer able to facilitate detoxication, leaving hepatocytes vulnerable to other xenobiotics as well as endogenous molecules.

Carbon Disulfide

Carbon disulfide (CS_2)is used in the production of cellophane and semiconductors and, less frequently, as a pesticide. Inhalation is the route of absorption, and biotrans-formation produces sulfur-containing metabolites that are eliminated in the urine. Distributed to the brain, CS_2 produces severe CNS and PNS toxicities, including organic brain damage, sleep disturbances, memory loss, Parkinson's disease-like symptoms, and ocular and auditory disorders.

Glycols

Of the glycols (ethylene, diethylene, and propylene), ethylene glycol is commonly known as automobile "antifreeze." Because of its sweet taste, ethylene glycol is often ingested by cats, dogs, and other domestic animals. Ethylene glycol undergoes a series of biotransformation reactions in the liver to form oxalic acid. In the kidney, oxalic acid accumulates in cells that form the tubules of the nephron and as crystalline precipitates within the tubules. Tubule cytotoxicity and the presence of crystals within the tubule are thought to contribute to renal failure. Propylene glycol antifreeze offers a less toxic alternative to traditional ethylene glycol, since it is biotransformed to lactic acid, pyruvic acid, and eventually CO_2 and water as a result of the Krebs cycle.

Aromatic Hydrocarbons

Named for their "aromatic" scent, these hydrocarbons each contain a six-carbon ring: benzene, a single ring, and alkylbenzene, a ring with a side chain (aliphatic). Used as a solvent and chemical feedstock, benzene has now replaced alkyl lead compounds as an "antiknocking" agent in fuels. Benzene is absorbed percutaneously or via the respiratory system. Toxicity likely results from benzene metabolites rather than directly from benzene. Hematotoxicity may lead to pancytopenia, as well as leukemia. Toluene, xylene, and ethylbenzene are examples of alkylbenzenes. Inhalation is the primary route of entrance, and toxicity predominately involves CNS depression.

Other Environmental Toxicants

Ionizing radiation in the form of alpha particles, beta particles, gamma rays, and x rays are usually not readily recognized as "toxicants." However, the energy contained in these agents is sufficient to damage cells and produce toxicity. During the early 1900s thousands of people were exposed to radium, either as participants in faddish medical therapies or as radium dial painters who "sharpened" the tips of small paint brushes by twirling the bristles on their tongue. Epidemiological studies of these two populations show an increased incidence of osteogenic carcinoma (bone cancer), probably the result of the substitution of radioactive ^{226}Ra for Ca during biomineralization.

Epidemiological studies of persons exposed to radioactive ^{224}Ra as a treatment for tuberculosis, x rays as a treatment for ringworm (both early to mid 1900s), and atomic bomb survivors (1945), also reveal these forms of ionizing radiation to be powerful genotoxic carcinogens. Dramatic increases in childhood thyroid cancer rates in countries most contaminated by the 1986 nuclear accident at Chernobyl have been linked to radiation exposure—most likely radioactive iodine released during the catastrophe (Figure 9-3).

Several epidemiological studies have linked exposure to electromagnetic fields (EMFs), associated with electrical transmission lines and household electrical wires/devices, with an increased risk of developing cancer. Two types of "fields," magnetic and electric, are of potential concern. However, researchers have been unable to determine if EMFs are causative or associative—no mechanism has been identified to explain the link between EMFs and suspected biological effects.

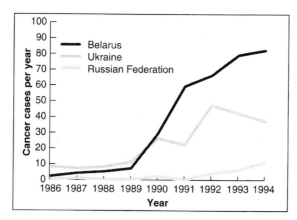

Figure 9-3. Increase in thyroid cancer cases per year following the nuclear accident in 1986 at Chernobyl. Source: WHO, 1995.

Review Questions

1.   Historical data suggest which category to be the greatest contributor to toxicity?

A.   Household products
B.   Industrial chemicals
C.   Man-made products
D.   Naturally occurring substances
E.   Pesticides

2.   Which of the following chemical categories are insecticides?

A.   Bipyridyls
B.   Carbamates
C.   Organochlorines
D.   A and B
E.   B and C

3.   On distribution to the lung, this toxicant affects gas exchange by damaging pneumocytes.

A.   Aldicarb
B.   Carbaryl
C.   DDT
D.   DNP
E.   Paraquat

4.   Which are true statements about Agent Orange?

A.   It is a 50:50 mixture of 2,4-D and 2,4,5-T.

B. It contains a trace impurity called dioxin or TCDD.
C. It was once marketed as an antiobesity agent.
D. A and B
E. A, B, and C

5. This toxicant has cataractogenic properties:

A. Ethylene-bisdithiocarbamate (EBDC)
B. 2,4-Dinitrophenol (DNP)
C. Fluoroacetic acid
D. Hexachlorobenzene (HCB)
E. Warfarin

6. Toxicity mechanisms of rodenticides include all of the following except:

A. Anticoagulants
B. Diabetogenics
C. Inhibitors of cellular respiration
D. Proximal tubule damage
E. Vasoconstrictors

7. Which are true statements about metals?

A. Chelating agents facilitate the removal of toxic metals from the body.
B. Some metals are essential to normal cellular functioning.
C. Metals are often destroyed by the human body.
D. A and B
E. A, B, and C

8. Sign(s) of chronic arsenic toxicity include:

A. Acute tubular necrosis
B. Cancer of the nasal cavities
C. Mee's lines
D. A and B
E. A, B, and C

9. In general, which toxic responses result from exposure to organic solvents?

A. Depression of the CNS
B. Irritation of tissues and membranes
C. Osteoporosis
D. A and B
E. A, B, and C

10. Discuss epidemiological studies that link ionizing radiation to toxicity.

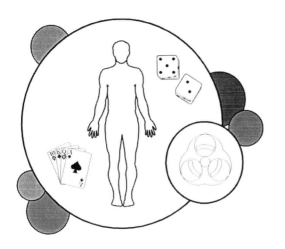

O bjectives

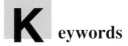

K eywords

- Define risk and safety

- Describe the use of the terms probability and incidence as related to risk

- Identify factors that contribute to differences in risk perception

- List the processes of risk assessment

- Summarize the parameters needed to estimate risk

- Recognize the importance of risk management

- Discuss the Safe Human Dose formula

- Explain the contributions of environmental toxicology to the survival of all organisms

exposure evaluation
incidence
lifetime average daily dose
   (LADD)
maximum daily dose (MDD)
probability
risk
risk assessment
risk estimation
risk management
risk perception
Safe Human Dose (SHD)
safety
Safety Factor (SF)
toxicant evaluation
toxicant identification

143

Introduction to Risk

Risk is defined as the *possibility* of loss or injury. Within the context of environmental toxicology this broad definition can be restated as the possibility of an undesirable biological response (toxicity) that results from exposure to a toxicant. Risk may be expressed as a probability (e.g., P = 0.00001) or incidence (e.g., 1 in 100,000) of a particular response for a given exposure. Risk (probability or incidence) is based on statistical estimates from sample populations studied during toxicity testing and on other observations, such as those from epidemiological studies. Risk statements, when properly interpreted, provide a valuable means for comparing the relative, but not necessarily absolute, possibility of a response occurring on exposure to a toxicant.

Each individual interprets risk in a unique way. Differences in risk perception may be attributed to how much a person knows about the toxicant, sources of exposure, and resulting toxicity. Two groups that frequently have *knowledge differences* at the core of their disagreements in risk perception are experts—often scientists—and lay persons (Table 10-1). Scientists have a social responsibility to educate the nonscientific community about the methods, results, and interpretations of toxicity tests and epidemiological studies regarding risks of environmental toxicants.

Also important to the perception of risk by individuals is their choice or control over exposure to a toxicant, their view of its catastrophic potential, and their concept of equanimity in the distribution of benefits and risks of the toxicant. Significant global differences, often delineated by geopolitical boundaries, are most likely influenced by cultural, social, and political factors.

Risk Assessment

Risk assessment is the process of examining toxicological and epidemiological data for a suspected toxicant and then, if warranted, estimating permissible exposures. Four steps characterize the process: (1) toxicant identification, (2) toxicant evaluation, (3) exposure evaluation, and (4) risk estimation.

Table 10-1. Differences in risk perception between experts and lay persons

Activity	Experts	Lay persons
Motor vehicles	1	2
Smoking	2	4
Alcoholic beverages	3	6
Handguns	4	3
Surgery	5	10
Motorcycles	6	5
X rays	7	22
Pesticides	8	9
Electric power (nonnuclear)	9	18
Swimming	10	19

Note. Environment, vol. 21, p. 14, 1979. Adapted with permission of the Helen Dwight Reid Educational Foundation. Published by Heldref Publications, 1319 Eighteenth St., NW, Washington, DC 20036-1802. Copyright © 1979.

These four steps may appear to have readily quantifiable, empirically derived answers—but not so! This is because science as a methodology is very good at proposing and answering questions about the natural world (Step 1), but science does not tell an environmental toxicologist how to interpret conclusions or what to do with answers in Steps 2–4. This presents a serious challenge to those involved in risk assessment, as human lives may ultimately be jeopardized.

Toxicant Identification

A review of existing literature, which may include toxicity tests and epidemiology, may be used to identify a toxicant. In the absence of relevant literature, descriptive toxicity testing must be done. Remember, the dose-response conclusions of descriptive toxicology are sufficient for toxicant identification—it is not necessary to understand the mode of action of a toxicant (mechanistic toxicology). The bottom line of toxicant identification is: Does the agent cause the adverse effect?

Toxicant Evaluation

Since toxicity tests are carried out in nonhuman species, the applicability of those tests to humans must be evaluated. An awareness of interspecific (between species) and intraspecific (within species) variability in toxicity testing is necessary when evaluating nonhuman tests. Careful attention to variability in age, sex, diet, circadian rhythms, hormonal status, and biotransformation capacity is required.

Questions to be asked may include: What type of test was performed— should a chronic toxicity test have been used instead of an acute test, or was a test performed to determine teratogenic activity? What responses were measured—what is the possibility of observing these same responses in humans? Are the results scientifically valid—can the results be reproduced in another laboratory using the same methods of toxicity testing? Were there problems with the test methods, choice of test organisms, doses tested, or in the stated results or interpretation of results?

Exposure Evaluation

With good experimental design, exposure to a suspected toxicant is precisely regulated during toxicity testing. Descriptive toxicologists predetermine the exposure parameters, including the organisms exposed, route of entry, dose, frequency (how often), and duration (how long) of dosing. Under "field" conditions the regulation of these parameters can present an illusive element to the risk assessor—the inability to control the test organisms (young, old, male, female, or pregnant female), route of entry (percutaneous, respiratory or digestive system), dose (?mg/kg), frequency of dose (?mg/kg/?min, day, week, year, lifetime), or duration of dose (seconds, minutes, hours, days, years).

Often overlooked, possibly because of its complexity, exposure assessment needs to be pursued with the same diligence as toxicant identification and evaluation. The assessment should examine exposures currently experienced as well as those anticipated under different conditions. Exposure for noncarcinogenic toxicants is expressed as maximum daily dose (MDD) (mg/kg/day), whereas exposure to carcinogens is stated as lifetime average daily dose (LADD) (mg/kg/day/lifetime).

Risk Estimation

Risk is the probability of an undesirable biological response resulting from exposure to a toxicant. Estimating risk requires an integration of the toxicity conclusions (toxicant identification and dose-response evaluation) and exposure assessment (MDD or LADD).

Risk is approximated by the equation:

$$R = T \times E$$

where risk = R, toxicity = T, and exposure = E. However, since dose-response relationships are not linear (as in the characteristic sigmoidal graph lines), exposure is more accurately expressed as a function (f) as seen in the following equation:

$$R = T \times f(E)$$

This equation is the source of risk statements such as 1 in 1,000 individuals will develop toxicity (a specific disease) if exposed to a specific dose (MDD or LADD) of a toxicant for a certain period of time.

Risk Management

Risk is often presented as a declarative statement, devoid of interpretation. To be of value to humanity, risk—once characterized—must be implemented into regulatory policies that benefit society. The intent of risk management is to examine risk assessment data and, where needed, develop regulatory options that address public health and social and economic concerns. Federal agencies charged with overseeing risk management often must overcome the influence of negative public perceptions and legislative mandates to arrive at responsible decisions.

A "healthy" approach to determining acceptable risk is to answer the following questions: Is the substance really needed? Could alternate, less toxic substitutes be used? What is the realistic amount of public exposure? What are the risks versus benefits for continued use of the agent? What is the environmental impact of the substance? Does the procurement of the agent deplete an environmental resource? Does existing technology permit the "final" disposal of the substance? If used, do we have the technology to ensure the "safe" use of the substance?

Safety

Safety is defined as the possibility that an undesirable biological response (toxicity) *will not result* from exposure to a toxicant—it is the inverse of the probability of risk (i.e., $1 \div P$). For example, when the risk of toxicity from a given exposure is $P = 0.00001$, safety equals 1 divided by 0.00001 (or 1/0.00001). It can be concluded that for every 100,000 exposures only 1 of those exposures will result in an adverse response.

Risk and safety are numerical estimates that result from consideration of toxicity and exposure—only two of the many factors influencing the capacity of a toxic agent to cause disease and death. Recall that toxicokinetic processes include absorption, distribution, storage, biotransformation, and elimination—all capable of altering the fate of a toxicant in the body. Also, since toxicity studies depend on nonhuman test organisms, some means of extrapolating results to humans is required. One calculation that permits more realistic extrapolations of toxicokinetic data from test organisms to humans is the Safe Human Dose (SHD) formula:

$$SHD = mg/day$$
$$= \frac{ED_{0.0} \times A_{t/h} \times T_{1/2_{t/h}} \times Wt \times D_{t/h}}{SF}$$

where

$ED_{0.0}$ = threshold dose of toxicant (NOEL)

$A_{t/h}$ = ratio of absorption of toxicant in test organism and human

$T_{1/2_{t/h}}$ = ratio of half-life of toxicant in test organism and human

Wt = weight of exposed individual

$D_{t/h}$ = ratio of toxicant test dosages in animals to exposure dosages in humans

SF = safety factor

Unlike the "measured" expressions in the numerator of the SHD equation, the safety factor (SF) in the denominator depends on the reliability of data used for extrapolation—a subjective judgment based on previous experience. The SF may range from 10 to 1,000, with lower SFs being used when valid human data is available and higher SFs indicative of a lack of human data. The goal of the SHD is to establish doses at which risk equals or approaches zero (P = 0.0).

Conclusions

The role of environmental toxicology is not only to identify environmental toxicants and their mode of action, but to assist in the evaluation and determination of issues concerning acceptable risk and safety—safe doses for humans and other species. Moreover, environmental toxicology data should prompt us to restrict or prohibit the use of agents toxic to plants and animals in our ecosystem. The fate of all species, including our own, depends on our ability to recognize and effectively control the "ripples" generated by environmental toxicants.

Review Questions

1. All of the following are true statements about risk, except:

A. It is defined as the possibility of loss or injury.
B. It may be expressed as probability or incidence.
C. It is based on statistical estimates from sample populations.
D. It is so absolute that there is only one interpretation.
E. It provides a valuable means of comparing the relative possibility of a response for two or more toxicants.

2. Risk perception is influenced by an individual's:

A. Knowledge of the toxicant
B. Ability to control exposure to the toxicant

C. View of the catastrophic potential of the toxicant
D. A and B
E. A, B, and C

3. Risk assessment includes all of the following, except:

A. Toxicant identification
B. Identification of toxicodynamics
C. Evaluation of toxicant
D. Exposure evaluation
E. Risk estimation

4. Science, as a methodology, is very good at telling environmental toxicologists what to do with the answers resulting from risk assessment.

A. True
B. False

5. Exposure assessment:

A. May include realistic assessment of "field" conditions as related to exposure.
B. Should examine exposures currently experienced as well as those anticipated under different circumstances.
C. Needs to be pursued with the same diligence as toxicant identification and evaluation.
D. A and B
E. A, B, and C

6. Which equation most accurately approximates risk?

A. $R = T/E$
B. $R = T \times E$
C. $R = T - f(E)$
D. $R = 1/P$
E. $R = T \times f(E)$

7. What questions need to be answered in order to pursue a healthy approach to determining acceptable risk?

8. Write out the Safe Human Dose equation and then discuss the objectivity of each term.

9. Discuss toxicant evaluation questions that you would ask as part of your approach to risk assessment.

10. What is the role of an environmental toxicologist?

Appendix

Resources in Environmental Toxicology

Books

Burrell , R. 1992. *Toxicology of the Immune System*. New York: Van Nostrand Reinhold.

Cockerham, L. G., & Shane, B. S. (Eds.). 1993. *Basic Environmental Toxicology*. Boca Raton, FL: CRC Press.

Davey, B., & Halliday, T. (Eds.). 1994. *Human Biology and Health: An Evolutionary Approach*. Buckingham, United Kingdom: Open University Press.

DiPiro, J. T., et al. (Eds.). 1993. *Pharmacotherapy: A Pathophysiologic Approach* (2nd ed.). Norwalk, CT: Appleton & Lange.

Emsley, J. 1994. *The Consumer's Good Chemical Guide: A Jargon-Free Guide to Controversial Chemicals*. New York: W. H. Freeman.

Francis, B. 1994. *Toxic Substances in the Environment*. New York: Wiley-Interscience.

Guthrie, F., & Perry, J. (Eds.). 1980. *Introduction to Environmental Toxicology*. New York: Elsevier.

Halstead, B. W. 1965. *Poisonous and Venomous Marine Animals of the World* (Vol. 1–3). Washington, DC: U.S. Government Printing Office.

Hayes, A. W. (Ed.). 1994. *Principles and Methods of Toxicology*. New York: Raven Press.

Hodgson, E., & Levi, P. E. 1994. *Introduction to Biochemical Toxicology* (2nd ed.). Norwalk, CT: Appleton & Lange.

Katzung, B. G. (Ed.). 1992. *Basic & Clinical Pharmacology* (5th ed.). Norwalk, CT: Appleton & Lange.

Klaassen, C. D., Amdur, M. O., & Doull, J. (Eds.). 1996. *Casarett and Doull's Toxicology: The Basic Science of Poisons* (5th ed.). New York: McGraw Hill.

Kupchella, C. E., & Hyland, M. C. 1989. *Environmental Science: Living Within the System of Nature* (2nd ed.). Boston, MA: Allyn and Bacon.

Landis, W. G. 1995. *Environmental Toxicology*. Boca Raton, FL: Lewis.

Lu, F. C. 1996. *Basic Toxicology: Fundamentals, Target Organs, and Risk Assessment* (3rd ed.). Washington, DC: Taylor & Francis.

O'Flaherty, E. J. 1981. *Toxicants and Drugs: Kinetics and Dynamics*. New York: Wiley.

Rubin, E., & Farber, J. L. (Eds.). 1994. *Pathology* (2nd ed.). Philadelphia, PA: J. B. Lippincott.

Smith, R. P. 1992. *A Primer of Environmental Toxicology*. Malvern, PA: Lea & Febiger.

Stacey, H. H. 1993. *Occupational Toxicology*. London, United Kingdom: Taylor & Francis.

Timbrell, J. A. 1991. *Principles of Biochemical Toxicology* (2nd ed.). London, United Kingdom: Taylor & Francis.

Timbrell, J. A. 1995. *Introduction to Toxicology* (2nd ed.). London, United Kingdom: Taylor & Francis.

Williams, P. L., & Burson, J. L. (Eds.). 1985. *Industrial Toxicology: Safety and Health Applications in the Workplace*. New York: Van Nostrand Reinhold.

Zakrzewski, S. F. 1991. *Principles of Environmental Toxicology*. Washington, DC: American Chemical Society.

Journals

Annual Review of Pharmacology and Toxicology (annually). Palo Alto, CA: Annual Reviews.

Chemical Research in Toxicology (8 per annum). Washington, DC: American Chemical Society.

Critical Reviews in Toxicology (bimonthly). Boca Raton, FL: CRC Press.

Drug and Chemical Toxicology (quarterly). New York: Marcel Dekker.

Drug Safety (bimonthly). Langhorne, PA: ADIS International.

Environmental Toxicology and Water Quality (quarterly). New York: Wiley.

Food and Chemical Toxicology (monthly). Oxford, United Kingdom: Pergamon Press.

Fundamental and Applied Toxicology (10 per annum). Orlando, FL: Academic Press.

Human and Experimental Toxicology (bimonthly). Hampshire, United Kingdom: Macmillan Magazines Ltd.

Inhalation Toxicology (9 per annum). Washington, DC: Taylor & Francis.

Journal of Analytical Toxicology (7 per annum). Niles, IL: Preston.

Journal of Applied Toxicology (bimonthly). West Sussex, United Kingdom: Wiley.

Journal of Toxicology and Environmental Health (18 per annum). Washington, DC: Taylor & Francis.

Journal of Toxicology–Clinical Toxicology (bimonthly). New York: Marcel Dekker.

Pharmacology & Toxicology (12 per annum). Copenhagen, Denmark: Munksgaard International.

The Toxicologist: An Official Publication of the Society of Toxicology (24 per annum). Reston, VA: Society of Toxicology.

Toxic Substance Mechanisms (quarterly). Washington, DC: Taylor & Francis.

Toxicology & Ecotoxicology News (bimonthly). London, United Kingdom: Taylor & Francis.

Toxicology and Applied Pharmacology (monthly). Orlando, FL: Academic Press.

Toxicology and Industrial Health (6 per annum). Princeton, NJ: Princeton Scientific.
Toxicology Letters (18 per annum). Ireland: Elsevier Science Ireland Ltd.
Xenobiotica (monthly). London, United Kingdom: Taylor & Francis.

Electronic Information Resources

Numerous toxicology-related databases are available on disk or CD-ROM and online. Hundreds of Internet resources may be accessed by a using a variety of software tools, called clients, including the following: mail (join a list), FTP (file transfer protocol), News (Usenet news groups), Gopher (for access to universities and governmental agencies), and the WWW (World Wide Web).

Glossary

A

absorption The taking in of substances by cells or tissues.

active transport The movement of molecules against a concentration gradient, from a less concentrated region to a more concentrated region.

acute toxicity The sudden onset of adverse health effects that are of a short duration; results in cellular changes that are reversible.

adenosine triphosphate (ATP) Energy-storing compound found in all cells.

adipose tissue A connective tissue composed primarily of adipocytes; functions in the storage of fat.

agenesis Absence or imperfect development of any body part.

Agent Orange An herbicide composed of a 50 : 50 mixture of 2,4-D and 2,4,5-T.

agranulocytopenia An acute condition characterized by a reduction in the number of monocytes and lymphocytes (i.e., agranular leukocytes).

albumin A simple protein distributed throughout the tissues and fluids of plants and animals.

aliphatic alcohols A class of organic compounds characterized by a straight- or branched-chain structure with an attached hydroxyl group.

alkylbenzenes A class of organic compounds containing a single benzene ring with one or more aliphatic side chains.

allergic contact dermatitis A delayed hypersensitivity reaction affecting the skin that results from exposure to a chemical.

alveolar region The area of the respiratory system that contains respiratory bronchioles and their associated alveoli.

Ames assay A bacterial test system used to determine mutagenicity.

anabolism Synthesis reactions in which smaller molecules are bonded together to form larger molecules; the reactions require energy and are catalyzed by enzymes.

anemia A deficiency of red blood cells or hemoglobin.

aneuploidy State of having an abnormal number of chromosomes not an exact multiple of the haploid number; presence of one more or one less chromosome.

anthracosilicosis The accumulation of carbon and silica in the lungs from inhaled coal dust that produces fibrous nodules.

anticoagulants Agents that prevent the aggregation of platelets; commonly used in rodenticides.

anticodon A triplet of bases in tRNA that matches a codon in mRNA.

antidotes Agents that neutralize or counteract the effects of a poison.

anuria The absence of urine formation.

aromatic hydrocarbons A class of unsaturated cyclic hydrocarbons containing one or more rings.

153

arterial blood gas (ABG) The concentration of oxygen and carbon dioxide in arteries.

arterial vessels Blood vessels that take blood away from the heart toward the capillaries.

asbestosis A lung disease resulting from inhalation of asbestos particles; sometimes complicated by mesothelioma.

associative relationship An epidemiological finding that establishes a link between two or more variables and a disease state.

atmosphere The region of the earth that contains gases (air).

atresia Absence or closure of a normal opening or normally continuous lumen.

B

base analogues Molecules whose chemical structure closely resembles pyrimidines and purines.

base substitution The exchange of a base (adenine, guanine, cytosine, thymine, or uracil) for a normally occurring base in the genetic code.

benzene A toxic hydrocarbon used as the basic structure in aromatic compounds, as a solvent, and as a chemical feedstock.

berylliosis A lung disease characterized by fibrosis that results from the inhalation of beryllium.

bioactivation A sequence of chemical reactions that produce intermediate or final metabolites that are more toxic or reactive than the original parent chemical; same as toxication.

biological half-life ($T_{1/2}$) The time required to reduce by half the quantity of a toxicant present in the body.

biological magnification The accumulation of toxicants or other chemicals by successive organisms in the food chain.

biosphere The region of the earth where life exists, including parts of the lithosphere, hydrosphere, and atmosphere.

biotransformation The process by which endogenous or exogenous substances are changed from hydrophobic to hydrophilic molecules to facilitate elimination from the body.

bipyridyls A class of chemical compounds characterized by two nitrogen-containing rings; examples include the herbicides diquat and paraquat.

blood dyscrasias Blood disorders that result from abnormal cellular components, such as too many of one blood cell type or too few of another.

blood flow/mass ratio A ratio of the volume of blood flowing through an organ to the size or mass of the organ.

blood plasma The yellowish, noncellular fluid portion of whole blood.

blood urea nitrogen (BUN) A test used to evaluate kidney function in which the concentration of urea in the blood is measured.

blood-brain barrier The barrier between the circulating blood and brain tissue formed by astrocytes and capillaries; prevents harmful substances in the blood from damaging neurons.

bronchoscopy The process of inserting a bronchoscope (small fiber-optic instrument) into the tracheobronchial region to visually examine the bronchi.

C

cancer The uncontrolled proliferation of cells.

capillaries Blood vessels that take blood from small arteries to small veins.

carbamates A class of compounds containing carbamic acid; as insecticides they are cholinesterase inhibitors.

carbon disulfide A toxic liquid used as an organic solvent and in the manufacture of rayon, cellophane, carbon tetrachloride, and rubber accelerators.

carbon tetrachloride A colorless toxic liquid used as an organic solvent; once widely used as a dry cleaning agent.

carcinogenesis The formation of cancer, including carcinomas and other malignant neoplasms.

carcinogens Cancer-producing substances.

cardiac output The volume of blood pumped per heart beat (stroke volume) times the heart rate (beats per minute); the resting average is 5 to 6 L/min.

catabolism Degradation reactions in which larger molecules are broken down to form smaller molecules; these reactions often release energy and are catalyzed by enzymes.

cataractogenic Cataract producing.

causal relationships A direct cause-and-effect linkage involving a single variable that is the basis for establishing a dose-response relationship.

ceiling effect The region (right side) on a cumulative dose-response graph where the line becomes almost horizontal indicating little or no change in response with increased doses.

cell membrane The membrane composed of phospholipids, proteins, and cholesterol that forms the outer boundary of a cell and regulates the movement of substances into and out of the cell.

cells The basic units of structure and function in a living organism; complex assemblages of atoms, molecules, and complex molecules.

cellular division The reproduction of somatic and germ cells.

central nervous system (CNS) The part of the nervous system that consists of the brain and spinal cord.

centromere The region on a chromosome where chromatids join together.

chelating agents Substances that facilitate the elimination of toxic metals by forming a metal–ion complex that is more reactive and more readily eliminated from the body.

chlorinated aliphatics A class of organic compounds characterized by a straight- or branched-chain structure with an attached chlorine group.

chloroform Trichloromethane, used as a solvent and to produce general anesthesia.

chlorophenoxy compounds A class of chemicals characterized by a phenol with attached chlorines; included are the herbicides 2,4-D and 2,4,5-T.

chromosome Structure made of DNA and protein found in the nucleus of a cell; human somatic cells contain 46 chromosomes.

chronic toxicity Adverse health effects that are of a long and continuous duration; due to irreversible cellular changes in the organism.

clinical toxicology The branch of toxicology that examines the effects of toxicants on an individual and the efficacy of treatment for symptoms related to toxication.

codon A sequence of three bases in DNA or mRNA that codes for one amino acid; also called a triplet code.

concentration gradient The relative amounts of a substance on either side of a membrane; diffusion occurs from the region of high concentration to the region of low concentration.

conjugate A metabolite that results from the joining of a phase II molecule with a toxicant (or its intermediate metabolite) that is more water-soluble than the original toxicant (or its intermediate metabolite).

conjugation reactions Phase II biotransformations in which a molecule provided by the body is added to a toxicant (or phase I metabolite).

creatinine A waste product produced when creatine phosphate is used for energy; excreted by the kidneys in urine, and often measured as an indicator of kidney function.

cumulative dose-response graph The cumulative sum of responses from lower to higher doses; the line on this graph appears sigmoidal.

cytochrome P-450 An iron–protein complex with a maximum absorbance of visible light at 450 nm that functions as a nonspecific enzyme system during phase I biotransformation reactions.

cytosolic enzymes Enzymes that are non-membrane-bound and occur free within the cytoplasm; catalyze phase II biotransformation reactions.

D

delayed toxicity The development of disease states or symptoms many months or years after exposure to a toxicant.

deoxyribonucleic acid (DNA) A nucleic acid with the shape of a double helix that is present in chromosomes; the repository of hereditary characteristics (genetic code).

dermatotoxicity The adverse effects produced by toxicants in the skin.

descriptive toxicology The branch of toxicology concerned with the identification of toxic substances.

detoxication The process by which toxicants are chemically converted to metabolites that are more readily eliminated by the urinary and biliary systems; same as detoxification.

developmental syndromes Multiple, but related, tissue or organ anomalies that may result from a teratogen.

diabetogenic Causing diabetes.

digestive system The organ system responsible for changing food into simple molecules that can be absorbed by the blood and lymph, and used by cells; made up of the digestive tract and related accessory organs (liver and pancreas).

dioxin A trace impurity (TCDD) associated with 2,4,5-T; a powerful toxicant.

diploid The normal number of chromosomes found in a somatic cell; in humans 2n = 46.

distribution A toxicokinetic process that occurs after absorption when toxicants enter the lymph or blood supply for transport to other regions of the body.

dithiocarbamates Chemical agents used as fungicides and insecticides.

division failures These result when, under the influence of teratogens, fused structures fail to separate (e.g., syndactyly).

dose-response relationship Exists when a consistent mathematical relationship describes the proportion of test organisms responding to a specific dose for a given exposure period.

dysraphic anomalies These result when, under the influence of teratogens, apposed structures fail to fuse (e.g., spina bifida).

E

Ebers papyrus One of eight Egyptian papyri, which dates from 1500 B.C., containing directions for the collection, preparation, and administration of more than 800 medicinal and poisonous recipes.

ecosystem A self-regulating community of animals and plants interacting with one another and with their nonliving environment.

ectopia This results when, under the possible influence of teratogens, organs or parts of the body are formed outside their normal location.

ED$_{50}$ The dose at which 50% of the test organisms are observed to exhibit an effective response.

effective dose (ED) A dose at which the predetermined response is observed.

efficacy The range of doses over which a toxicant produces a response; a toxicant is said to have a higher efficacy when the dose-response relationship continues to be present over a greater range of doses.

electromagnetic fields Fields of force that consist of associated electric and magnetic components; possess a specific amount of electromagnetic energy.

elimination The toxicokinetic processes responsible for the removal of toxicants or their metabolites from the body.

embryogenesis The formation of the embryo.

embryolethality Death of the fertilized ovum or embryo during the first 8 weeks (i.e., embryonic stage).

end effect A response that is observed and recorded during toxicity testing.

endocytosis The process, including pinocytosis and phagocytosis, whereby substances are taken into a cell by invagination of the cell membrane.

environmental toxicants Agents in our surroundings that are harmful to human health.

environmental toxicology The study of the poisons around us; the hazardous effects of poisons on human health.

epidemiology The study of the prevalence and spread of disease and death in a population.

epigenetic A carcinogenic mechanism that does not act to directly affect DNA, termed non-DNA reactive.

epithelium One of the four main types of tissues; forms glands, lines cavities, and covers body surfaces.

erythrocytes Red blood cells.

ethyl alcohol Ethanol or beverage alcohol.

ethylene glycol A thick, sweet, colorless liquid used as antifreeze, coolant, and hydraulic fluid.

etymology The study of word origins.

exocytosis The cellular secretion of macromolecules by the fusion of vesicles with the cell membrane; the process of transporting cellularly derived substances across the cell membrane.

exposure evaluation The process of determining the validity of exposures used in toxicity testing as compared to the anticipated or known exposures encountered under field conditions.

F

facilitated diffusion The spontaneous passage of molecules and ions that are bound to specific carrier proteins across the cell membrane; dependent on the concentration gradient.

fecal excretion The process by which toxicants or their metabolites enter bile for transport to the duodenum and subsequent elimination.

fetal alcohol syndrome (FAS) A collection of signs and symptoms (e.g., fetal malformation, intrauterine growth retardation, craniofacial anomalies, and CNS dysfunction) found in offspring of mothers who are chronic alcoholics.

fluoroacetic acid Two derivatives, sodium fluoroacetate and fluoroacetamide, are rodenticides that act to block the Krebs cycle, impeding a major pathway for ATP production.

forced vital capacity (FVC) A pulmonary function test that measures the time it takes to exhale the total volume of air contained in the lungs (i.e., inhalatory reserve capacity, tidal volume, and exhalatory reserve capacity).

forensic toxicology The branch of toxicology concerned with medical and legal questions relating to the harmful effects of known or suspected toxicants.

frameshift mutation A mutation occurring when the number of nucleotides deleted or inserted is not a multiple of three; produces an improper grouping of codons.

frequency dose-response graph A graph that plots the percentage of organisms (Y axis) responding to a given dose (X axis); usually recognized by its bell-shaped appearance.

fungicides Agents that destroy or repel fungi.

G

gene Hereditary unit; portion of the DNA on a chromosome that represents a sequence of bases that contains the "blueprint" for a protein.

genetic code The sequence of three nucleotides (codon) that signifies a specific amino acid; there are four nucleotides that in different combinations of three result in 64 possible codons.

genotoxic A carcinogenic mechanism that acts directly to affect DNA, termed DNA reactive.

germ cells Egg or sperm cells; ova or spermatozoa.

glial cells Support cells (nonconducting) in the nervous system; include astrocytes, microglia, and oligodendrocytes.

glomerular filtration The first process in urine formation; results when blood enters the vascularized glomerulus where water and small molecules are forced by hydrostatic pressure across the glomerular filter and into Bowman's capsule.

glucuronidation The process of adding glucuronide to a toxicant or phase I metabolite during phase II biotransformation.

glycols Compounds containing adjacent alcohol groups; ethylene glycol is the simplest glycol.

glycosuria The presence of sugar in the urine.

H

haploid Denoting the number of chromosomes in sperm or ova; in humans, n = 23.

hazardous waste Waste that, because of its biological, chemical, or physical characteristics, or quantity or concentration, may produce disease.

hematotoxicity The presence of disease in the blood as produced by a toxicant.

hematotoxins Agents that produce toxic symptoms (i.e., disease) in the blood.

hematuria The presence of blood cells in the urine.

hemolytic anemias A reduction in the oxygen-carrying capacity of blood resulting from the destruction of erythrocytes.

hepatotoxicity The presence of disease in the liver as produced by a toxicant.

hepatotoxins Agents that produce disease in the liver.

herbicides Agents that destroy or inhibit plant growth.

hexachlorobenzene A chemical once used to prevent fungal infestation in seeds before planting.

histogenesis The formation of tissues in the body.

hydrolysis The chemical process whereby a compound is split into two or more simpler compounds with the uptake of H and OH parts of a water molecule on either side of the chemical bond at the splitting site.

hydrophilic Attracted to water.

hydrophobic Repelled by water.

hydrosphere That portion of the earth composed of water.

hypoplasia Underdevelopment of a tissue or an organ that results from a decrease in the number of cells.

hypoxia A decrease in the normal level of oxygen in the blood; inadequate supply of oxygen to the tissues.

immediate toxicity The rapid occurrence of symptoms following exposure to a toxicant.

in vitro In an artificial environment such as a test tube or tissue culture medium; not in a living organism.

in vivo Processes or reactions occurring within a living organism.

incidence A statistical estimate of the risk of a particular response for a given exposure (e.g., 1 in 100,000).

industrial toxicology The branch of toxicology concerned with the study of toxic symptoms (i.e., diseases) found in individuals who have been exposed to toxicants in their place of work.

infinite dilution The disposal of "small" quantities of wastes in "large" repositories such as the hydrosphere, lithosphere, and atmosphere where they are thought to pose little harm because the repositories are so vast.

initiation The subtle alteration of DNA or proteins within target cells by carcinogens that renders the cell capable of becoming cancerous.

insecticides Agents that destroy or repel insects.

integumentary system Composed of skin and related structures such as hair, glands, and nails; largest organ system in the human body.

interstitial fluid Fluid found between cells; intercellular.

intestinal excretion Process by which toxicants or their metabolites are transported from the blood in the submucosal region into the intestinal lumen.

intracellular fluid Fluid found within a cell; cytoplasm.

inulin A fructose polysaccharide used by intravenous injection to determine the rate of glomerular filtration.

ionization Occurs when atoms or molecules dissociate into electrically charged atoms or molecules; to be ionized.

ionizing radiation Energy radiated in the form of waves or particles, such as alpha particles, beta particles, gamma rays, and x rays.

irritant contact dermatitis Inflammatory response that occurs in skin on exposure to a toxicant.

karyotype A systematized array of chromosomes arranged in pairs in descending order of size and position of the centromere.

L

LD_{50} The dose at which 50% of the test organisms are observed to exhibit a lethal response.

lethal dose (LD) The dose that results in the death of the test organism.

leukemia Blood disease characterized by an increase in white blood cells.

leukocytes White blood cells.

lifetime average daily dose (LADD) Allowable exposure to carcinogens; expressed as mg/kg/day/lifetime.

linear dose sequence A series of test doses that increase by adding a constant; for example 1, 2, 3, 4, 5, 6, 7, 8, 9, and 10 mg/kg where the constant is 1.

lipid soluble Substances that are dissolved by nonpolar solvents or by the lipid region of the phospholipid bilayer that forms the cell membrane.

lipophilic Attractive to lipid-soluble substances.

lithiosphere The region of the earth's surface composed of rock.

local toxicity Occurs when symptoms are restricted to the site of initial exposure to a toxicant.

logarithmic dose sequence A series of test doses that increase by multiplying each previous dose by 10; for example, 0.1, 1.0, 10, 100, 1,000 and 10,000 mg/kg.

lymph Watery fluid in the lymphatic system.

lymph capillaries Microscopic vessels that transport lymph.

lymph node A rounded mass of lymphoid tissue surrounded by a capsule of connective tissue; lymph vessels enter and exit lymph nodes.

lymphatic system Part of the circulatory system that drains excess fluid from the tissues; includes lymph capillaries, lymph vessels, lymph nodes, aggregations of lymphoid tissue such as the tonsils, spleen, and thymus, and circulating lymphocytes.

lymphocyte One of the five different types of white blood cells; agranular white blood cells that function in immunity.

lymphoid tissue Present in the tonsils, spleen, and thymus.

M

macromolecules Large molecules.

margin of safety Expresses the magnitude of the range of doses between a noneffective or minimally effective dose and a lethal dose; expressed as a ratio such as LD_{50}/ED_{50}.

maximum daily dose (MDD) Allowable exposure to noncarcinogenic toxicants; expressed as mg/kg/day.

mechanistic toxicology The branch of toxicology concerned with determining the biochemical processes by which identified toxic substances have an impact on the organism.

Mee's lines Horizontal white bands seen in the nails of persons experiencing chronic arsenic toxicity.

meiosis A special process of cell division that halves the chromosome number in the formation of gametes.

metabolism The sum of biochemical changes occurring in an organism; includes anabolism, those reactions that convert small molecules into large, and catabolism, those reactions that convert large molecules into small.

metaphase A stage of mitosis (or meiosis) in which the chromosomes are aligned on the equatorial plate of the cell with the centromeres mutually repelling each other.

methyl alcohol Wood alcohol or methanol.

microcytic hypochromic anemia Reduced oxygen-carrying capacity that results when red blood cells are unusually small and lack the normal quantity of hemoglobin.

microsomal enzymes Enzymes associated with phase I biotransformation reactions; although normally bound to the endoplasmic reticulum, upon disruption of cells, small spherical vesicles (microsomes) derived from the endoplasmic reticulum contain the enzymes.

minute volume respiration (MVR) The amount of air breathed in on each respiratory cycle multiplied by the number of respirations per minute; expressed as liters per minute (L/min).

mitosis A type of cell division that results in the production of two daughter cells identical in chromosome number as compared to the parent cell.

mixed toxicity This results when one toxicant is not consistently more potent over the range of doses tested as compared to another toxicant; evident when two lines on cumulative dose-response graphs cross one another; same as reversed toxicity.

molecule Particle formed by chemical bonding of two or more atoms; smallest subunit of a compound.

monosomy Occurs when a cell loses one member of a pair of chromosomes; in humans, the presence of 45 chromosomes or $2n - 1$.

morbidity A diseased state.

morphogenesis The differentiation of cells and tissues in the embryo which results in establishing the form of organs and parts of the body.

mortality Death.

mucociliary escalator Responsible for the movement of mucus from the respiratory system; involves the coordinated sweeping action of cilia found on the surface of columnar epithelial cells that line the trachea, bronchi, and bronchioles.

mutagenesis The production of a mutation or change in the genetic code.

mutagens Agents that cause the production of mutations.

myelin Coiled fatty membrane that covers and insulates the axons of some neurons.

N

nasopharyngeal region Region of the respiratory system composed of the nares, nasal cavity, and adjacent pharynx.

nephritic syndrome Clinical symptoms characterized by hematuria or blood cells in the urine.

nephron The functional unit of the kidney; composed of the Bowman's capsule, proximal tubule, and distal tubule.

nephrotic syndrome Clinical symptoms characterized by proteinuria or protein in the urine.

nephrotoxicity The adverse effects produced by toxicants in the kidney.

nerve Collection of individual motor or sensory neurons.

neurons Nerve cells; composed of the cell body, dendrites, and axon.

neurotoxicity The adverse effects produced by toxicants in the nervous system.

neurotransmitter A chemical released from the distal axonal region of a presynaptic neuron that enables a nerve impulse to cross a synaptic junction to stimulate or inhibit a postsynaptic neuron; acetylcholine.

nickel itch Allergic contact dermatitis that results from exposure to nickel in sensitized individuals.

No Observable Effect Level (NOEL) A subthreshold dose; $ED_{0.0}$.

nonbiodegradable Resistant to biological degradation.

nonphotodegradable Resistant to degradation from ultraviolet radiation.

nonpolar molecules Chemicals that possess no positive or negative molecular charge; lipid-soluble.

normal distribution Represented by a bell-shaped line on a frequency dose-response graph or a sigmoidal line on a cumulative dose-response graph; the word "normal" should not be interpreted as "usual" or as the opposite of "abnormal."

nucleic acids Biological molecules (DNA and RNA) that permit organisms to reproduce; composed of purines or pyrimidines, a sugar, and a phosphate; a sequence of nucleotides.

nucleotide The building block of nucleic acids (i.e., DNA and RNA), consisting of a five-carbon sugar bonded to a nitrogen-containing base (adenine, cytosine, guanine, thymine, or uracil) and a phosphate group.

O

obstructive uropathies Renal diseases that result when the flow of urine is blocked by intra- or extratubular pathologies.

occluding cell junctions The tight junctions formed between adjacent cells; prevent the entrance of substances by blocking their passage along cell–cell junctions.

oliguria Scanty or small amount of urine production.

oogenesis The process in the ovary that results in the production of female gametes.

organ system Two or more organs that together perform a special function.

organelle One of many subcellular units that performs a specific function in the cell.

organochlorines A category of insecticides composed of chlorinated hydrocarbons; DDT, dieldrin, and Kepone.

organogenesis The formation and development of organs in the body.

organomercurials A category of fungicides that includes methylmercury.

organophosphates A category of insecticides that includes parathion, diazinon, and malathion.

organs A group of several tissue types that unite to form structures that perform a special function.

Overton's Rules Permeability is directly proportional to the lipid solubility of the toxicant and inversely proportional to the molecular size of the toxicant.

oxidation The loss of hydrogen or electrons by a compound or element; the reverse of reduction.

P

PAH clearance A kidney function test that examines the capacity of nephrons to remove *p*-aminohippuric acid from blood plasma; used to estimate the rate of plasma flow through the kidneys.

pancytopenia A reduction in the number of all blood cell types.

partition coefficient The ratio of a toxicant's solubility in a nonpolar solvent to its solubility in water.

percutaneous absorption The passage of substances through unbroken skin.

peripheral nervous system (PNS) The sensory and motor neurons that connect to the central nervous system; all the nerves and nervous tissue outside the central nervous system.

pesticides Agents that act to destroy or repel pests.

phagocytosis The process by which large particles are taken into a cell; cellular eating.

phase I biotransformation A chemical reaction that exposes or adds a small polar group to a toxicant; enhances the water solubility of a toxicant.

phase II biotransformation A chemical reaction that adds a large molecule to a reactive site on a toxicant or its metabolite to enhance water solubility.

phospholipid Molecule containing phosphates and lipids found in the cell membrane; the phosphate head is a region that is hydrophilic, whereas the lipid tail is a region that is hydrophobic.

phospholipid bilayer Term used to describe the "sandwich" appearance of the cell membrane; two layers of phospholipid molecules in which the phosphate heads are found on the inner and outer surfaces and the lipid tails are directed inward.

phototoxicity A form of dermatotoxicity that results when skin is overexposed to ultraviolet light or from the combination of specific wavelengths of light and a phototoxic substance.

phthalimides A category of potent fungicides that includes Captafol and Folpet; similar in structure to thalidomide.

phytotoxin A poison produced by a plant.

pinocytosis An active transport mechanism used to transfer fluids or dissolved substances into cells; cellular drinking.

placental barrier A barrier that limits the diffusion of water-soluble toxicants from maternal to fetal circulation.

plasma protein A simple protein distributed in the fluids of organisms; albumin is the most abundant circulating plasma protein to which toxicants are bound.

platelets The smallest cellular components of blood; thrombocytes.

pneumoconioses Abnormal conditions resulting from mineral dust in the lung; commonly leads to pulmonary fibrosis.

pneumocytes Special cells that line the alveoli of the lung.

poietins Stimulating factors that regulate the fate of stem cells in the production of blood components; erythropoietins.

point mutation A change in the chromosome at a single nucleotide within a gene.

poisons Substances that in relatively small doses act to destroy life or seriously impair cellular function.

polar molecules Chemicals that possess a positive or negative molecular charge; water-soluble.

polyploidy A chromosomal aberration in which there are more than two complete chromosome sets; more than 2n, such as 3n, 4n, or 5n.

portal vein The major vein entering the liver.

potency The toxicological activity of a substance; a relative concept for comparing toxicants.

potent A toxicant's capacity to effect a response; when comparing two toxicants, the one with the smaller ED_{50}, TD_{50}, or LD_{50} is the more potent.

probability A statistical estimate of the risk of a particular response for a given exposure to a toxicant (e.g., $P = 0.00001$).

procarcinogen A nonreactive substance that by itself cannot lead to the formation of cancer, but once activated can become a carcinogen.

promotion Occurs when initiated cells are acted upon by promoting agents to give rise to cancer.

proteinuria The presence of protein in the urine.

pulmonary fibrosis The abnormal condition of fibrous tissue as a reparative or reactive process in the lung.

pulmonotoxicity The adverse effects produced by toxicants in the respiratory system.

purines Nucleotides formed from two carbon–nitrogen rings; adenine and guanine.

pyrimidines Nucleotides formed from a single carbon–nitrogen ring; cytosine, thymine, and uracil.

Pyrinimil A rodenticide that interferes with glucose metabolism; diabetogenic.

R

reabsorption The second process in urine formation; results in the return to the blood of water, glucose, potassium, and amino acids lost during filtration.

reduction The gain of hydrogen or electrons by a compound or element; the reverse of oxidation.

regulatory toxicology The branch of toxicology concerned with assessing the data from descriptive and mechanistic toxicology to determine the legal uses of specific chemicals and the risks posed to the ecosystem by their marketing.

replication The process of duplicating a cell's DNA so that on cellular division (mitosis) the genetic content of each daughter cell will be identical to the parent cell.

respiratory system The organ system that functions as an air distributor and gas exchanger; made up of the nares, pharynx, larynx, trachea, bronchi, and lungs.

reversed toxicity Results when one toxicant is not consistently more potent over the range of doses tested as compared to another toxicant; evident when two lines on cumulative dose-response graphs cross one another; same as mixed toxicity.

ribonucleic acid (RNA) A single-stranded nucleic acid involved in protein synthesis, the structure of which is determined by DNA; mRNA is the end result of transcription, and rRNA, tRNA, and mRNA are all involved in translation.

risk The possibility of loss or injury.

risk assessment The process of examining toxicological and epidemiological data and estimating permissible exposures.

risk estimation The integration of toxicant identification and exposure evaluation to estimate risk.

risk management The use of risk assessment conclusions in the development and implementation of regulatory options that address public health, social, and economic concerns.

risk perception How each individual interprets the possibility of loss or injury.

rodenticides Agents that destroy or repel rodents.

S

Safe Human Dose (SHD) A formula used to extrapolate toxicokinetic data from test organisms to humans; expressed as mg/day.

safety The possibility that an undesirable biological response (toxicity) will not result from exposure to a toxicant; the inverse of the probability of risk.

Safety Factor (SF) A subjective value in the denominator in the Safe Human Dose formula that reflects the uncertainties inherent in extrapolating toxicokinetic data from test organisms to humans; a small value (e.g., 10) indicates that valid human data is available, and a large value (e.g., 1,000) indicates a lack of relevant human data.

secretion The third process in urine formation that occurs in the distal convoluted tubules; involves the active transport of molecules out of the blood and into the urine.

semipermeable membrane A characteristic of cell membranes; permits the passage of some molecules but not others.

simple diffusion The spontaneous movement of a substance down its concentration gradient from a more concentrated to a less concentrated region.

slow vital capacity (SVC) A measurement of the total volume of air contained in the lungs.

somatic cell Any cell in a multicellular organism other than a germ cell (sperm or egg).

spermatogenesis The process of meiosis in the testes to produce sperm cells.

storage The accumulation of toxicants or their metabolites in specific tissues or as bound to circulating plasma proteins.

subthreshold doses Doses at which no response is observed as a result of toxicity testing; NOEL.

sulfate conjugation A phase II biotransformation reaction in which phosphoadenosyl phosphosulfate (PAPS) reacts with a toxicant or phase I metabolite to produce a polar sulfate conjugate.

systemic toxicity Occurs when toxicants are absorbed at one site and distributed to a distant body region where they produce adverse effects.

T

target organ toxicity The adverse effects or disease states manifested in specific organs of the body.

TD$_{50}$ The dose at which 50% of the test organisms are observed to exhibit a toxic response.

teratogen An agent that alters normal cellular differentiation or growth processes, which results in a malformed fetus.

teratogenesis The origin or production of a malformed fetus.

teratology The branch of science concerned with the production, development, and classification of malformed fetuses; study of developmental anomalies in fetuses.

thermoplastics Polymers that can be shaped by pressure and heat to the form of a mold; they can also be remelted and remolded.

thermosetting plastics Polymers that can be shaped by pressure and heat to the form of a mold; they cannot be remelted and remolded.

Threshold dose The dose at which the first response is observed as a result of toxicity testing; below this dose no responses are observed.

Threshold Limit Value (TLV) Doses workers can be exposed to for 8 hours per day, 5 days per week for a working lifetime without exhibiting adverse health effects.

thrombocytopenia A condition in which there is an abnormally small number of platelets in the blood.

tissue A collection of cells that together perform a similar function.

toxic dose (TD) The dose at which test organisms are observed to exhibit toxicity.

toxicant An agent capable of producing symptoms of intoxication or poisoning.

toxicant evaluation Process of evaluating the applicability of results from toxicity testing performed on nonhuman organisms to humans.

toxicant identification Process whereby existing literature is reviewed to identify a toxicant.

toxication A sequence of chemical reactions producing intermediate or final metabolites that are more toxic or reactive than the original parent chemical: same as bioactivation.

toxicity The state of being poisonous; the adverse effects or symptoms produced by toxicants or poisons in organisms.

toxicity testing The scientific methodology used to establish a dose and response relationship.

toxicodynamics Study of the mechanisms by which toxicants produce their unique effects in an organism.

toxicokinetics Study of time-dependent processes related to toxicants as they interact with living organisms; includes study of absorption, distribution, storage, biotransformation, and elimination.

toxins Poisonous substances produced by plants, animals, or bacteria.

tracheobronchial region Region of the respiratory system composed of the trachea, bronchi, and bronchioles.

transcription Process of transferring genetic information from a DNA molecule into an RNA molecule.

translation Process of transferring information from an RNA molecule into a protein.

trisomy Occurs when a cell gains one member of a pair of chromosomes; in humans, the presence of 47 chromosomes or $2n + 1$.

vasoconstrictors Agents that initiate the contraction of the smooth muscles surrounding peripheral blood vessels.

venoms Poisonous substances secreted by certain animals, such as bees, spiders, and snakes.

venous vessels Blood vessels that take blood away from the capillaries toward the heart.

volume of distribution (V_D) The volume of body fluids into which a toxicant is distributed; plasma, interstitial fluid, and cellular fluid.

warfarin An anticoagulant that prevents the aggregation of thrombocytes; used as a rodenticide.

xenobiotics Substances not naturally produced within an organism; substances foreign to an organism.

Index